SpringerWienNewYork

Fortschritte der Chemie
organischer Naturstoffe

Progress in the Chemistry
of Organic Natural Products

Founded by L. Zechmeister

Editors:
A. D. Kinghorn, Columbus, OH
H. Falk, Linz
J. Kobayashi, Sapporo

Honorary Editor:
W. Herz, Tallahassee, FL

Editorial Board:
V. Dirsch, Vienna
S. Gibbons, London
N. H. Oberlies, Research Triangle Park, NC
Y. Ye, Shanghai

89

Fortschritte der Chemie
organischer Naturstoffe

Progress in the Chemistry
of Organic Natural Products

Authors:
B. Kräutler
N. P. Sahu, S. Banerjee, N. B. Mondal,
and D. Mandal

SpringerWienNewYork

Prof. A. Douglas Kinghorn, College of Pharmacy,
Ohio State University, Columbus, OH, USA

em. Univ.-Prof. Dr. H. Falk, Institut für Organische Chemie,
Johannes-Kepler-Universität, Linz, Austria

Prof. Dr. J. Kobayashi, Graduate School of Pharmaceutical Sciences,
Hokkaido University, Sapporo, Japan

This work is subject to copyright.
All rights are reserved, whether the whole or part of the material is concerned, specifically those of translation, reprinting, re-use of illustrations, broadcasting, reproduction by photocopying machines or similar means, and storage in data banks.

© 2008 Springer-Verlag/Wien
Printed in Germany

SpringerWienNewYork is part of
Springer Science + Business Media
springer.at

Product Liability: The publisher can give no guarantee for the information contained in this book. This also refers to that on drug dosage and application thereof. In each individual case the respective user must check the accuracy of the information given by consulting other pharmaceutical literature. The use of registered names, trademarks, etc. in this publication does not imply, even in the absence of a specific statement, that such names are exempt from the relevant protective laws and regulations and therefore free for general use.

Library of Congress Catalog Card Number AC 39-1015

Typesetting: Thomson Digital, Chennai
Printing and binding: Strauss GmbH, 69509 Mörlenbach, Germany

Printed on acid-free and chlorine-free bleached paper
SPIN: 12100125

With 7 partly coloured Figures

ISSN 0071-7886
ISBN 978-3-211-74018-7 SpringerWienNewYork

Contents

List of Contributors..................................... VII

Chlorophyll Catabolites
B. Kräutler... 1

1. Introduction.. 2
2. Chlorophyll Catabolites from Vascular Plants......... 6
   2.1. Green Chlorophyll Degradation Products in Vascular Plants.......... 6
       2.1.1. Chlorophyllide *a* and *b* from Chlorophylls by Loss
              of the Phytol Side Chain........................... 6
       2.1.2. Reductive Path from *b*- to *a*-Type Chlorophyll(ide)s........... 8
       2.1.3. Pheophorbide *a* from Chlorophyllide *a* by Removal
              of the Magnesium Ion............................... 8
       2.1.4. 13^2-Carboxy-pyropheophorbide *a* from Hydrolysis and
              Pyropheophorbide *a* from Overall Loss of the Methoxycarbonyl
              Group from Pheophorbide *a*......................... 9
   2.2. Non-green Chlorophyll Degradation Products from Vascular Plants..... 10
       2.2.1. Discovery and Structure Analysis of Fluorescent Chlorophyll
              Catabolites.. 11
       2.2.2. Preparation of the Elusive Red Chlorophyll Catabolite by Partial
              Synthesis.. 13
       2.2.3. An Enzyme-bound Red Chlorophyll Catabolite from Enzymatic
              Oxygenation of Pheophorbide *a*..................... 16
       2.2.4. Fluorescent Chlorophyll Catabolites from Enzymatic Reduction
              of the Red Chlorophyll Catabolite 18
       2.2.5. Model Experiments for the Reduction of the Red Chlorophyll
              Catabolite to Fluorescent Chlorophyll Catabolites........... 19
       2.2.6. Non-fluorescent Colourless Chlorophyll Catabolites 21
       2.2.7. A Non-enzymatic Tautomerization Achieves the "Final"
              Transformation of Fluorescent Chlorophyll Catabolites to
              Non-fluorescent Colourless Chlorophyll Catabolites 22
       2.2.8. Peripheral Functional Groups and Conjugations Found
              in Non-fluorescent Colourless Chlorophyll Catabolites 24
       2.2.9. Evidence for Further Breakdown of the Non-fluorescent
              Colourless Chlorophyll Catabolites in Higher Plants........... 28
3. Chlorophyll Catabolites from the Green Alga *Chlorella protothecoides*...... 30
4. Chlorophyll Catabolites from Marine Organisms 32

5. Conclusions and Outlook.................................... 34

Acknowledgements... 37

References... 37

Steroidal Saponins
N. P. Sahu, S. Banerjee, N. B. Mondal, and *D. Mandal*................... 45

1. Introduction... 45
2. Isolation.. 46
3. Structure Elucidation.................................... 49
   3.1. Conventional Methods............................... 50
   3.2. Spectrometry Coupled with Chemical Methods.......... 52
   3.3. Modern Spectrometric Methods....................... 55
      3.3.1. Mass Spectrometry............................. 55
      3.3.2. NMR Spectroscopy.............................. 57
         3.3.2.1. ^{1}H NMR Spectroscopy................... 57
         3.3.2.2. ^{13}C NMR Spectroscopy................ 58
         3.3.2.3. 2D NMR Spectroscopy...................... 59
4. Biological Activity..................................... 62
   4.1. Cytotoxic Activity Against Cancer Cell Lines........ 63
   4.2. Antifungal Activity................................ 66
   4.3. Miscellaneous Effects.............................. 68
5. Biosynthesis of Steroidal Glycosides.................... 69
6. Report of New Steroidal Saponins (1998–Mid-2006)........ 70
7. Conclusion.. 126

Acknowledgement... 126

References... 127

Author Index... 143

Subject Index.. 153

Listed in Medline

List of Contributors

Banerjee, Dr. S., Indian Institute of Chemical Biology, 4 Raja S C Mullick Road, Jadavpur, Kolkata 700 032, India

Kräutler, Prof. Dr. B., Institut für Organische Chemie, Universität Innsbruck, Innrain 52a, 6020 Innsbruck, Austria
e-mail: bernhard.kraeutler@uibk.ac.at

Mandal, Dr. D., Indian Institute of Chemical Biology, 4 Raja S C Mullick Road, Jadavpur, Kolkata 700 032, India

Mondal, Dr. N. B., Indian Institute of Chemical Biology, 4 Raja S C Mullick Road, Jadavpur, Kolkata 700 032, India

Sahu, Dr. N. P., Indian Institute of Chemical Biology, 4 Raja S C Mullick Road, Jadavpur, Kolkata 700 032, India
e-mail: npsahu@iicb.res.in

Chlorophyll Catabolites[*]

Bernhard Kräutler[**]

Institute of Organic Chemistry and Centre for Molecular Biosciences,
University of Innsbruck, 6020 Innsbruck, Austria

Contents

1. Introduction	2
2. Chlorophyll Catabolites from Vascular Plants	6
2.1. Green Chlorophyll Degradation Products in Vascular Plants	6
2.1.1. Chlorophyllide *a* and *b* from Chlorophylls by Loss of the Phytol Side Chain	6
2.1.2. Reductive Path from *b*- to *a*-Type Chlorophyll(ide)s	8
2.1.3. Pheophorbide *a* from Chlorophyllide *a* by Removal of the Magnesium Ion	8
2.1.4. 13^2-Carboxy-pyropheophorbide *a* from Hydrolysis and Pyropheophorbide *a* from Overall Loss of the Methoxycarbonyl Group from Pheophorbide *a*	9
2.2. Non-green Chlorophyll Degradation Products from Vascular Plants	10
2.2.1. Discovery and Structure Analysis of Fluorescent Chlorophyll Catabolites	11
2.2.2. Preparation of the Elusive Red Chlorophyll Catabolite by Partial Synthesis	13
2.2.3. An Enzyme-bound Red Chlorophyll Catabolite from Enzymatic Oxygenation of Pheophorbide *a*	16
2.2.4. Fluorescent Chlorophyll Catabolites from Enzymatic Reduction of the Red Chlorophyll Catabolite	18
2.2.5. Model Experiments for the Reduction of the Red Chlorophyll Catabolite to Fluorescent Chlorophyll Catabolites	19
2.2.6. Non-fluorescent Colourless Chlorophyll Catabolites	21
2.2.7. A Non-enzymatic Tautomerization Achieves the "Final" Transformation of Fluorescent Chlorophyll Catabolites to Non-fluorescent Colourless Chlorophyll Catabolites	22
2.2.8. Peripheral Functional Groups and Conjugations Found in Non-fluorescent Colourless Chlorophyll Catabolites	24
2.2.9. Evidence for Further Breakdown of the Non-fluorescent Colourless Chlorophyll Catabolites in Higher Plants	28

---

[*] Dedicated to the memory of my mother, Prof. Margarethe Kräutler, Teacher of Nature's secrets.
[**] E-mail: bernhard.kraeutler@uibk.ac.at

3. Chlorophyll Catabolites from the Green Alga *Chlorella protothecoides* 30
4. Chlorophyll Catabolites from Marine Organisms 32
5. Conclusions and Outlook 34
Acknowledgements 37
References 37

1. Introduction

This chapter reviews the occurrence, structure, and reactivity of chlorophyll catabolites from vascular plants and from some microorganisms. In parallel, synthetic means for obtaining such tetrapyrrolic compounds are recapitulated. The available structural information on chlorophyll catabolites (*1*) has provided a basis for deriving much of the current insights into the biochemical pathways of chlorophyll breakdown in plants and for complementary plant-biological work, as has been reviewed elsewhere recently (see Scheme 1) (*2, 3, 4, 5, 6*).

Breakdown of the green plant pigments and the emergence of autumnal colours in the foliage of deciduous trees represent most fascinating natural phenomena (*7*) (see Fig. 1). In spite of the high visibility of these processes, in the early 1990s still, breakdown of chlorophyll in plants was considered to be an enigma (*8*). The plant chlorophylls (Chls), chlorophyll *a* (Chl *a*, **1a**) and chlorophyll *b* (Chl *b*, **1b**), even seemed to disappear "without leaving a trace" (*9*). The earlier search for Chl-catabolites was generally directed at finding coloured compounds and has remained rather fruitless: indeed, the first chlorophyll catabolites to be identified from higher plants turned out to be colourless tetrapyrroles (*10*).

Due to their unique roles in photosynthesis, the chlorophylls have a special position among the natural porphinoids (*11, 12, 13*). Indeed, biosynthesis and degradation of the green pigments are probably the most visual sign of life on earth (*8*), and are observable even from outer space (*4*) (see Fig. 1). Although considerable work has been done on the biosynthesis of the chlorophylls (*14, 15, 16*), there has been a definitive lack of information on the fate of the green plant pigments. According to recent estimates, more than 10^9 tons of chlorophyll (Chl) are biosynthesized and degraded every year on the earth (*8*). In view of the obvious ecological and economic relevance of these intriguing processes, the fate of Chl and Chl-breakdown are of considerable interest.

References, pp. 37–43

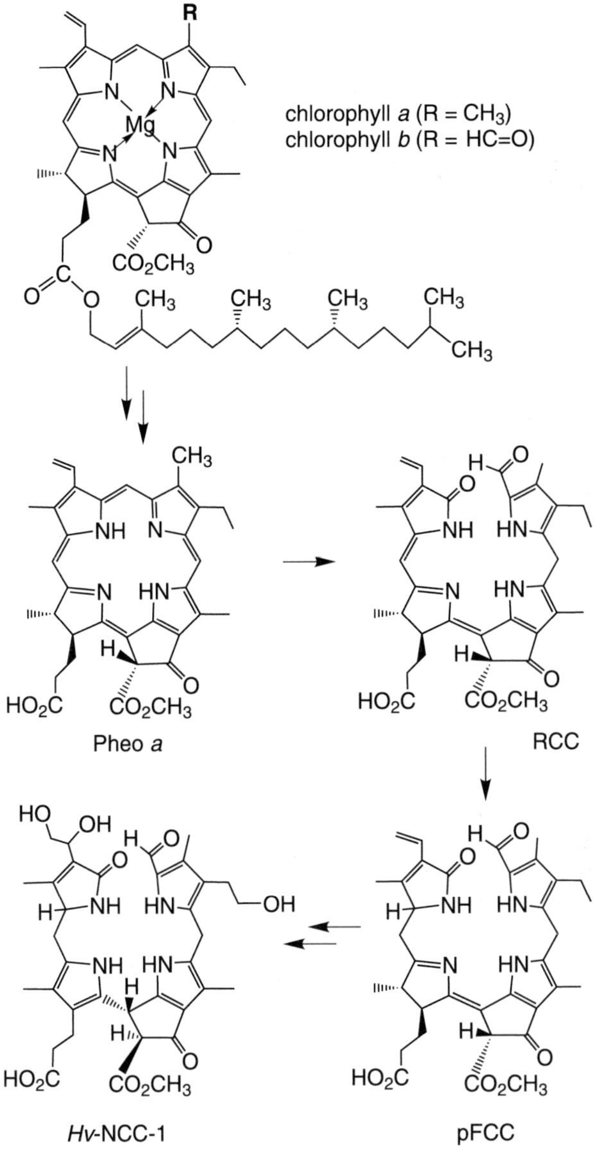

Scheme 1. Overview of chlorophyll breakdown in senescent higher plants (2). The chlorophylls (Chl *a*, **1a** (R = CH$_3$) or Chl *b*, **1b** (R = CH=O) are degraded *via* pheophorbide *a* (Pheo *a*, **5a**), "red" chlorophyll catabolite (RCC, **11**), the primary "fluorescent" chlorophyll catabolites (pFCCs, **10**) to "non-fluorescent" chlorophyll catabolites (NCCs), such as *Hv*-NCC-1 (**2**, also called RP-14)

Fig. 1. Top: Satellite images of Europe, colour coded according to the Vegetation Index and taken in June (left) and October 2000 (right) (made available by Deutsches Fernerkundungsdatenzentrum (DFD), Oberpfaffenhofen, Germany). Bottom: Senescent leaves of a Katsura tree (*Cercidiphyllum japonicum*) growing in the Hofgarten, Innsbruck, and pictured in October 2003

By analogy to heme breakdown in plants and animals (*17*), an oxygenolytic opening of the porphinoid macrocycle of the Chls was commonly considered as the key step in Chl-breakdown (*8*). Based on

References, pp. 37–43

experiences on the reactivity of chlorins towards electrophilic agents (*18*), it was assumed that an opening of the Chl-macrocycle would occur at the "western" δ-*meso* position, *i.e.* next to the peripherally reduced ring D of the macrocycle (*8*). Photo-oxygenolysis of chlorins indeed was found to preferentially occur at the δ-*meso* position and thus served as a chemical model (*19*). The structural analyses by Kishi and coworkers of luciferin from the dinoflagellate *Pyrocystis lunula* and of a luminescent compound from krill appeared to strengthen the relevance of this observation for Chl-breakdown (see formulae of compounds **33** and **34** in Section 4): Both of these compounds were found to be linear tetrapyrroles that were most likely derived from Chls by opening at the δ-position of the macro-ring (*20, 21*).

Studies by Matile and coworkers of senescent leaves of a non-de-greening genotype of the grass *Festuca pratensis* gave first good evidence for the existence of non-green Chl-catabolites in leaf extracts (*22, 23*): Comparison of the extracts from the senescent (de-greened) wild-type leaves with those from the non-de-greening mutant by analysis by thin-layer chromatography revealed the formation of pink and rust-coloured spots on the silica-gel plates, in the case of the wild-type leaves only. These coloured compounds were termed "pink pigments" and "rusty pigments" and were suggested to be chemical degradation products of what seemed to be colourless Chl-catabolites originally. Similar compounds were found in yellowing primary leaves of barley (*24, 25*), when forced to de-green in permanent darkness. Surprisingly they were found in the vacuoles, rather than in the de-greened chloroplasts, from where they must have originated (*24*). Incorporation of ^{14}C isotopic label from 4-^{14}C-δ-aminolevulinic acid suggested the role of Chls as the precursor of the "rusty pigments" (*26*). One of them, called "rusty pigment 14" originally (but later designated as *Hv*-NCC-1, **2**), was identified as a colourless catabolite of Chl *a* (**1a**) by spectroscopic means and its constitution could be established unambiguously as that of a $3^1,3^2,8^2$-trihydroxy-1,4,5,10,15,20-(*22H,24H*)-octahydro-13^2-(methoxycarbonyl)-4,5-dioxo-4,5-seco-phytoporphyrinate, see Scheme 2 and Section 2.2.6 below) (*10, 27*).

This work revealed the first structure of a non-green Chl-catabolite from plants and gave first-hand clues as to the major structural changes occurring in the degradation of Chl during senescence, as further discussed below. Indeed, the major Chl-catabolites from vascular plants are now known to have the same basic skeleton as **2** and to be colourless "non-fluorescent" chlorophyll catabolites (NCCs). The NCC-structures, such as of *Hv*-NCC-1 (**2**), were clearly incompatible with a catabolic relevance in Chl-breakdown of an oxygenolytic opening at the

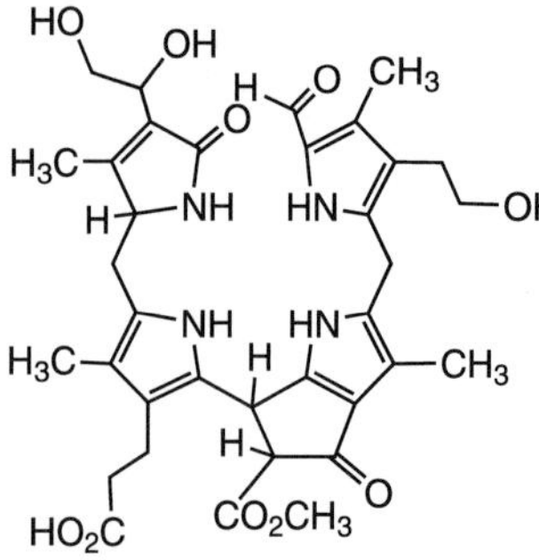

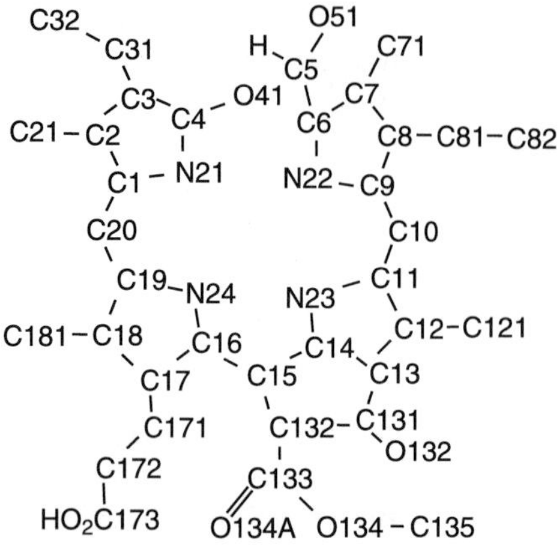

Scheme 2. Left: Constitutional formula of *Hv*-NCC-1 (**2**), originally named RP-14 (*10*); right: atom numbering used, which is based on the numbering of the Chls (*10*)

α-position of the chlorin macrocycle, as well as of some "early" hydroxylation reactions at the intact chlorin macrocycle of the Chls (*2, 10*). It was also remarkable to see that the genetic control of chlorophyll breakdown had a crucial impact on the development of the laws of genetics, which Mendel established in the last century (*28*). The puzzling observation of the phenotype of a recessive allele in Mendel's "green peas" is now known to be due to a specific gene involved in chlorophyll breakdown and recently identified in a variety of plants, including peas (*29*).

2. Chlorophyll Catabolites from Vascular Plants

2.1. Green Chlorophyll Degradation Products in Vascular Plants

2.1.1. Chlorophyllide a and b from Chlorophylls by Loss of the Phytol Side Chain

The structure of *Hv*-NCC-1 (**2**) was consistent with the loss of the phytol side chain from Chl *a* (**1a**) as an early event of Chl-breakdown. The enzymatic hydrolysis of Chl *a* (**1a**) to chlorophyllide *a* (**3a**) and to phytol by chlorophyllase was discovered in the early 20th century by A. Stoll (see Scheme 3) (*30*). Chlorophyllase removes the lipophilic phytol anchor of the Chl-molecules, which is crucial for binding of the

References, pp. 37–43

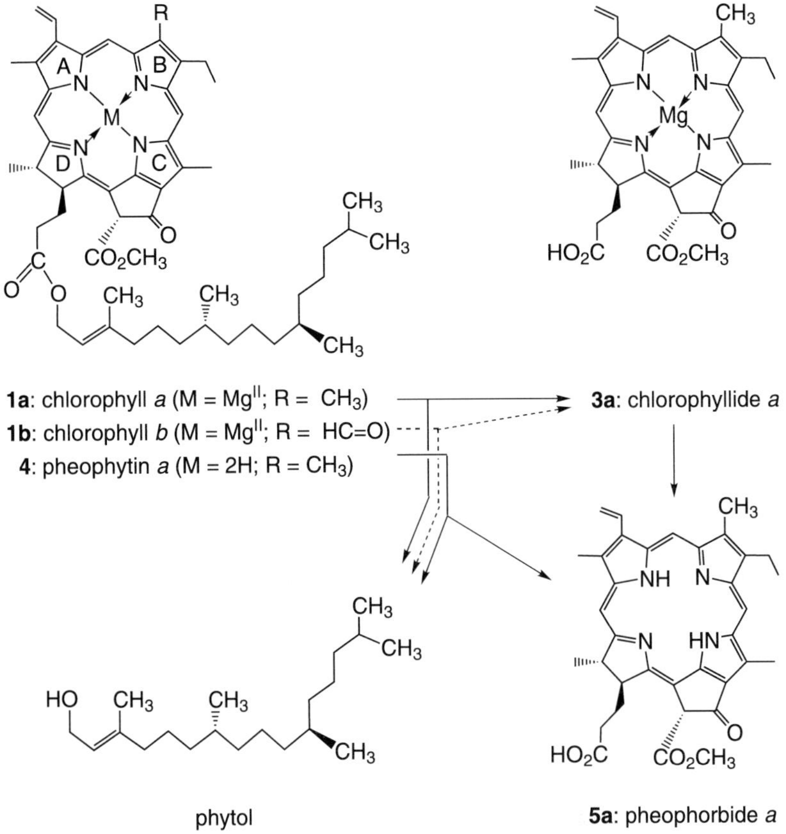

1a: chlorophyll a (M = Mg^{II}; R = CH$_3$)
1b: chlorophyll b (M = Mg^{II}; R = HC=O)
4: pheophytin a (M = 2H; R = CH$_3$)

3a: chlorophyllide a

phytol

5a: pheophorbide a

Scheme 3. Chlorophyll a (R = CH$_3$, **1a**) or chlorophyll b (R = CH=O, **1b**) are degraded via chlorophyllide a (**3a**) to pheophorbide a (**5a**) and phytol (recovered as phytyl-acetate); alternatively, pheophytin a (**4**) is also hydrolyzed to **5a** and phytol

green pigment to the Chl-binding proteins and for insertion of the Chl-protein complexes into the thylakoid membranes of chloroplasts (*31*). Chlorophyllase is localized in the chloroplast envelope (*32*) and hydrolyses or *trans*-esterifies not only Chl a (**1a**), Chl b (**1b**), but also pheophytin a (**4**) (*33*). Hydrolytic loss of phytol has recently been shown to set the stage for further enzymatic degradation of both the Chls and the proteins (*3, 34*). In the course of leaf senescence, the total content of phytol is remarkably constant: in de-greened barley leaves, it is stored as phytyl acetate in the lipid rich plastoglobuli of the senescent chloroplasts (*35*).

2.1.2. Reductive Path from b- to a-Type Chlorophyll(ide)s

The NCCs detected in extracts from senescent leaves of vascular plants were all found (with one exception (*36*), see Section 2.2.4) to have a 7-methyl group, as is present in Chl *a* (**1a**). The fate of the *b*-type Chls in Chl-breakdown was, therefore, a matter of particular interest (*3*). The absence of catabolites derived from Chl *b* (**1b**) was puzzling, at first. The finding of a biochemical pathway from the *b*-type to the *a*-type chlorophyll(ide)s helped to rationalize it (*15, 37, 38, 39*): chlorophyllide *b* (**3b**) is transformed to chlorophyllide *a* (**3a**) by reduction of the 7-formyl group of **3b** to a 7-methyl group (as in **3a**) in a sequence involving two enzymes (*15*).

The well-established biosynthetic oxidation of the *a*-type to the *b*-type Chls (*15, 40*) has thus obtained an unexpected reductive counterpart. The two counteracting redox sequences now represent a "(Chl *a*/Chl *b*)-cycle", which can help to regulate the (Chl *a*/Chl *b*)-ratio in plants for the purpose of adapting the photosynthetic apparatus to the light intensity (*15, 40*). Clearly, the reductive part has the additional role as a very early and obligatory step in chlorophyll breakdown (*15*). The reduction of chlorophyllide *b* (**3b**) to chlorophyllide *a* (**3a**) ensures that all the plant Chls are made available for the catabolic "pheophorbide *a*" pathway (*2, 3, 4, 5*). This is important, since the crucial and senescence specifically expressed oxygenase that cleaves the chlorin macrocycle accepts pheophorbide *a* (Pheo *a*, **5a**), but is inhibited by Pheo *b* (**5b**) (*41*) (see Section 2.2.2 below). When primary leaves of barley were artificially de-greened in the presence of deuterated water, the NCC *Hv*-NCC-1 (**2**) was found to carry a mono-deuterated 7-methyl group, consistent with the operation of the chlorophyll(ide) *b* (to *a*) reduction during Chl-catabolism (*42*).

2.1.3. Pheophorbide a from Chlorophyllide a by Removal of the Magnesium Ion

Most of the available information on Chl-breakdown suggested dephytylation and reductive conversion of *b*-chlorophyll(ide) to *a*-type analogues to precede the loss of the magnesium ion (*2, 3, 4, 5, 43*). Removal of the magnesium ion from chlorophyllide *a* (**3a**) occurs with extreme ease in dilute acid and generates Pheo *a* (**5a**). In senescent cotyledons of oilseed rape (*44*), as well as in leaves of *Chenopodium album* (*43*) activity of a magnesium dechelating enzyme has been observed. A large fraction of the magnesium set free by the degradation of chlorophyll during senescence is transported out of the senescent leaf and stored in the remaining part of the plant (*9*).

References, pp. 37–43

2.1.4. 13^2-Carboxy-pyropheophorbide a from Hydrolysis and Pyropheophorbide a from Overall Loss of the Methoxycarbonyl Group of Pheophorbide a

Pyropheophorbide a (Pyropheo a, **6**) was observed in *Chenopodium album* and was considered as an "early" catabolite of Chl-degradation in this green plant (*45*, *46*). A related study with *Chlamydomonas reinhardtii* gave results that supported this view (*47*): When senescence of this green alga was artificially induced by lack of light, while Chl-

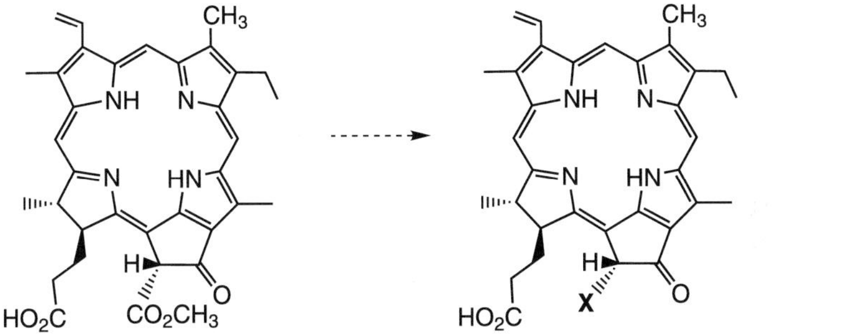

5a: pheophorbide a ⟶ **7**: 13^2-carboxy-pyropheophorbide a (X = CO_2H)
6: pyropheophorbide a (X = H)

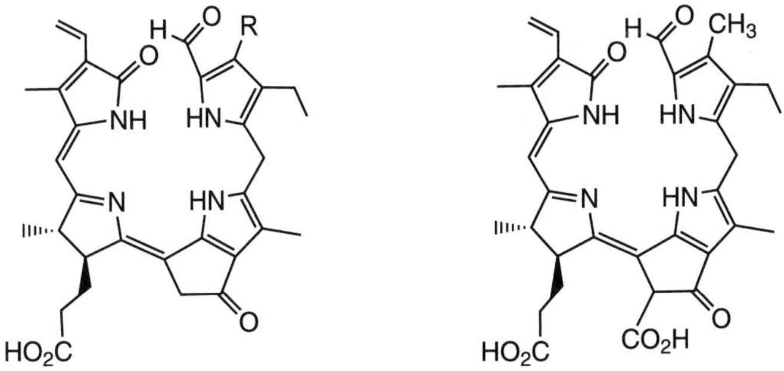

8a: R = CH_3
8b: R = CH=O

9

Scheme 4. In *Chenopodium album* pheophorbide a (**5a**) is degraded to pyropheophorbide a (**6**) via 13^2-carboxy-pyropheophorbide a (**7**). The red tetrapyrroles **8a** and **8b** were isolated from the culture medium of the green alga *Chlorella protothecoides* (the monoacid **8a** is likely to be a nonenzymatic decarboxylation product of the diacid **9**) (*58*, *59*)

degradation was blocked due to strictly anaerobic conditions, Pheo *a* (**5a**) and Pyropheo *a* (**6**) accumulated. However, it remains to be seen whether **6** represents an early intermediate of Chl-breakdown in senescing plants and algae: so far, a non-green tetrapyrrolic Chl-catabolite having a 13^2-methylene group (as in **6**) has not been isolated from senescent higher plants (*1, 2, 3, 4, 5, 36, 48, 49, 50, 51, 52, 53, 54, 55, 56*).

In *Chenopodium album* significant amounts of 13^2-carboxy-pyropheophorbide *a* (**7**) were identified, suggesting that only hydrolysis of the methyl ester function of Pheo *a* (**5a**) was enzyme-catalyzed (*43*). As expected (*48*), the β-ketocarboxylic acid function of **7** underwent non-enzymatic decarboxylation readily at ambient temperature to give Pyropheo *a* (**6**) (*43*), supporting the feasibility of a non-enzymatic origin of the latter (see Scheme 4). Related observations have been made with a red isolate from the green alga *Chlorella protothecoides* (*57*): When the origin of the red, ring opened derivative **8a** of Pyropheo *a* (**6**) from Chl-breakdown was reinvestigated, **8a** was found to be due to a non-enzymatic decarboxylation during work-up of the dicarboxylic acid **9** (with a β-keto-carboxylic acid function, see Scheme 16 in Section 3, below) (*58, 59*). Indeed, at present, all known natural NCCs from higher plants still carry either a methoxycarbonyl group or a carboxylic acid function at the crucial 13^2-position (*2*). A direct link between the observation of Pyropheo *a* (**6**) and the later stages of Chl-catabolism in higher plants (and green algae) is thus lacking (*2, 5*). However, the relevance for Chl-catabolism of the enzymatic hydrolysis of the 13^2-methoxy-carbonyl group of Pheo *a* (**5a**) (observed in *Chenopodium album* (*43*)) may not be discounted, as a variety of NCCs (and an FCC) were indicated to carry a 13^2-carboxyl functionality (see *e.g.* (*48, 52, 60*)).

2.2. Non-green Chlorophyll Degradation Products from Vascular Plants

Colourless, non-fluorescent Chl-catabolites (NCCs) have meanwhile been observed to accumulate in a variety of senescent vascular plants (*1, 2, 3, 4, 5, 36, 48, 49, 50, 51, 52, 53, 54, 55, 56*). All of them feature an annealed cyclopentanone unit, substituted by a carboxylate or methoxy-carbonyl function (*1*), a hallmark of the natural chlorophyll derivatives (*61*). The molecular constitution of the NCCs revealed an intriguing and specific oxygenolytic ring-opening reaction at the α-*meso* position (rather than at the δ-*meso* carbon) of the chlorin macrocycle with retention of the α-*meso* carbon as a formyl group (*1*).

References, pp. 37–43

2.2.1. Discovery and Structure Analysis of Fluorescent Chlorophyll Catabolites

The structures of the NCCs were (with one exception (*36*), see Section 2.2.4) consistent with a direct lineage to chlorophyll(ide) *a*. At the same time, their complex build-up indicated the involvement of several (enzymatic) steps in their formation from Pheo *a* (**5a**), their common precursor (*41*). In the context of the search for possible intermediates on the way to the NCCs, the fleeting appearance of nearly colourless but fluorescent compounds in senescent cotyledons of oilseed rape ("*Brassica napus*") was intriguing. These fluorescent compounds could be seen most clearly, when the apparent rates of Chl-breakdown were high (*62*). They were provisionally named "fluorescing Chl-catabolites" (FCCs), because ^{14}C-labeling identified them as porphyrin derivatives (*63, 64, 65*). As none of these fluorescent compounds accumulated *in vivo*, they were considered to represent precursors for the NCCs and possibly even the "primary" products of cleavage of the porphinoid macrocycle of **5a**. This assumption was strengthened by locating such fluorescent compounds in intact chloroplasts isolated from senescent leaves of barley, from where they were released under appropriate conditions (*64*). On the other hand, the long sought for discovery of coloured Chl-catabolites from senescent higher plants (*9*) was not achieved in a variety of related experiments (*63, 66*).

An extract of the chloroplast membranes from senescent cotyledons of oilseed rape eventually constituted an *in vitro* system for the preparation of a larger sample of an FCC. It contained the needed enzymatic oxygenating activity and converted Pheo *a* (**5a**) into about 5% of an FCC (isolated by HPL-chromatography) (*62*). The constitution of this apparently rather labile FCC (named *Bn*-FCC-2) was elucidated by mass spectrometric and NMR-spectroscopic means (*62*): the molecular formula of *Bn*-FCC-2 (**10**) was determined by high-resolution mass spectrometry as $C_{35}H_{40}N_4O_7$. Formally, this indicated **10** to differ from **5a** only by addition of one equivalent of molecular oxygen and two equivalents of molecular hydrogen. The NMR-derived constitution identified **10** as a $3^1,3^2$-didehydro-1,4,5,10,17,18,20-(22*H*)-octahydro-13^2-(methoxycarbonyl)-4,5-dioxo-4,5-seco-phytoporphyrin. *Bn*-FCC-2 (**10**) is a linear tetrapyrrole, derived from Pheo *a* (**5a**) by an oxygenolytic cleavage at the α-*meso* position and by saturation of the β- and δ-*meso* positions (*62*). Consistent with a chromophore, extending over rings C and D, the UV/Vis-spectrum of **10** now shows two prominent bands, near 361 and 320 nm (*1*). Aqueous solutions of the FCC **10** show strong luminescence, with a maximum near 436 nm, as was also

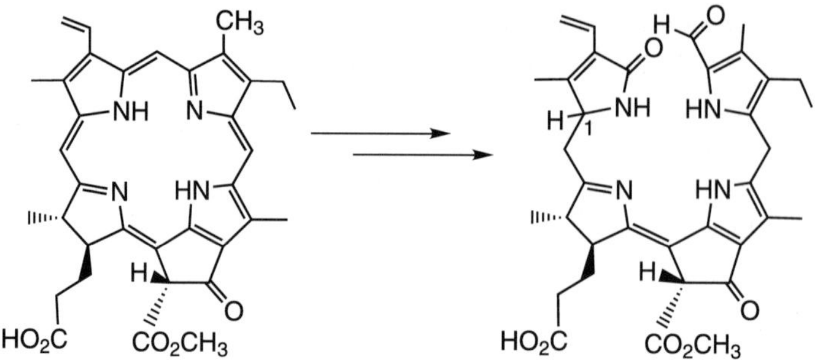

5a: pheophorbide *a*

"primary" fluorescent chlorophyll catabolites
10 (pFCC) and *epi*-**10** (*epi*-pFCC)

Scheme 5. In higher plants pheophorbide *a* (**5a**) is degraded to the "primary" fluorescent chlorophyll catabolite (pFCC, **10**) and to its C(1)-epimer *epi*-**10** (*epi*-pFCC)

observed for the fleetingly existing fluorescing compounds. The derived structure of *Bn*-FCC-2 (**10**) clearly identified it as an intermediate in Chl-breakdown preceding the stage of the NCCs: the characteristic complete de-conjugation of the four pyrrolic units of the tetrapyrrolic NCCs could result, formally, from the FCC **10** by a tautomerization reaction (see Section 2.2.4 below) (*62*). The constitution of the FCC **10** reflected the minimal transformations needed to convert green Pheo *a* (**5a**) into a colourless compound with the chromophore of an FCC (see Scheme 5). The fluorescent catabolite from oilseed rape, *Bn*-FCC-2 (**10**), therefore, was postulated to represent the first formed or "primary" FCC (or pFCC) (*4, 62*).

A second fluorescent Chl-catabolite, *Ca*-FCC-2, was isolated from another *in vitro* system, based on enzymatic activity obtained from ripe (red) sweet pepper (*Capsicum annuum*) and its structure was analyzed (*67*). The new "fluorescent" catabolite could be shown by mass spectrometry to be an isomer of **10**: Further NMR-spectroscopic analysis revealed *Ca*-FCC-2 to have the same constitution and to differ from pFCC (**10**) only in the absolute configuration at C(1). *Ca*-FCC-2 was thus assigned as the epimeric "primary" 1-*epi*-pFCC (*epi*-**10**) (*67*).

As is delineated in more detail below (Section 2.2.3), the chiral center C(1) is introduced *via* the highly stereo-selective reduction step catalyzed by a reductase, present in the two plant species (*67, 68, 69*). These findings, identified the two FCCs (**10** and *epi*-**10**) as direct products of these reductases and supported the earlier proposal to consider both of these fluorescent compounds as "primary" fluorescent Chl-catabolites (*2, 3*).

References, pp. 37–43

A further important piece of information about the early steps in Chl-breakdown was supplied by the discovery that Pheo *a* (**5a**) accumulated in the absence of molecular oxygen in the higher plant *Festuca pratensis* (*70*), but not Pheo *b* (**5b**). This finding suggested the involvement of both O_2 and **5a**, as common substrates in the oxidative enzymatic step during Chl-breakdown that cuts open the chlorin macrocycle. As described in the next section, an enzyme bound, ring-opened "red" chlorophyll catabolite (RCC) was indeed found to be the product of this oxygenase, which is now called "pheophorbide *a* oxygenase" (PaO) (*5*).

2.2.2. Preparation of the Elusive Red Chlorophyll Catabolite by Partial Synthesis

The structure of the "primary" fluorescent Chl-catabolite pFCC (**10**, $3^1,3^2$-didehydro-1,4,5,10,17,18,20-(22*H*)-octahydro-13^2-(methoxy-carbonyl)-4,5-dioxo-4,5-seco-phytoporphyrin, see Scheme 5) (*62*), and other findings (*41*, *71*), made the cleavage of the porphinoid macro-ring of Pheo *a* (**5a**) by an oxygenase a likely "key step" in Chl-breakdown (*2*, *72*). The putative oxygenase, whose activity depended upon an iron-containing reactive center (but not upon a heme cofactor) (*71*), was considered likely to be related to other non-heme iron-dependent (mono)-oxygenases. An oxygenolytic opening of the macro-ring at its α-*meso* position might give the elusive "red" tetrapyrrole **11**, which,

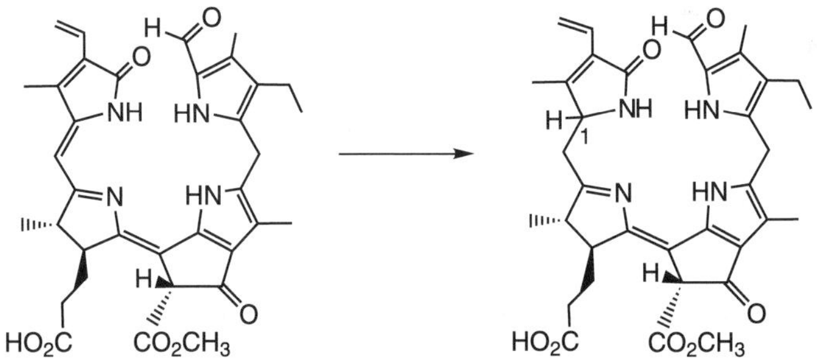

11: "red" chlorophyll catabolite (RCC)

10: "primary" fluorescent chlorophyll catabolite (pFCC)
epi-**10**: 1-*epi*-pFCC (C(1)-epimer of pFCC)

Scheme 6. "Primary" fluorescent catabolites (pFCCs) **10** and *epi*-**10** result from enzymatic reduction of the elusive red chlorophyll catabolite (RCC, **11**) by RCC-reductase

therefore, would represent a putative intermediate in chlorophyll breakdown (*62*). The red compound **11** was suggested to be, potentially, a direct precursor of **10**: a reduction step, involving the addition of two hydrogen atoms of the "western" δ-*meso* position and at C(1),

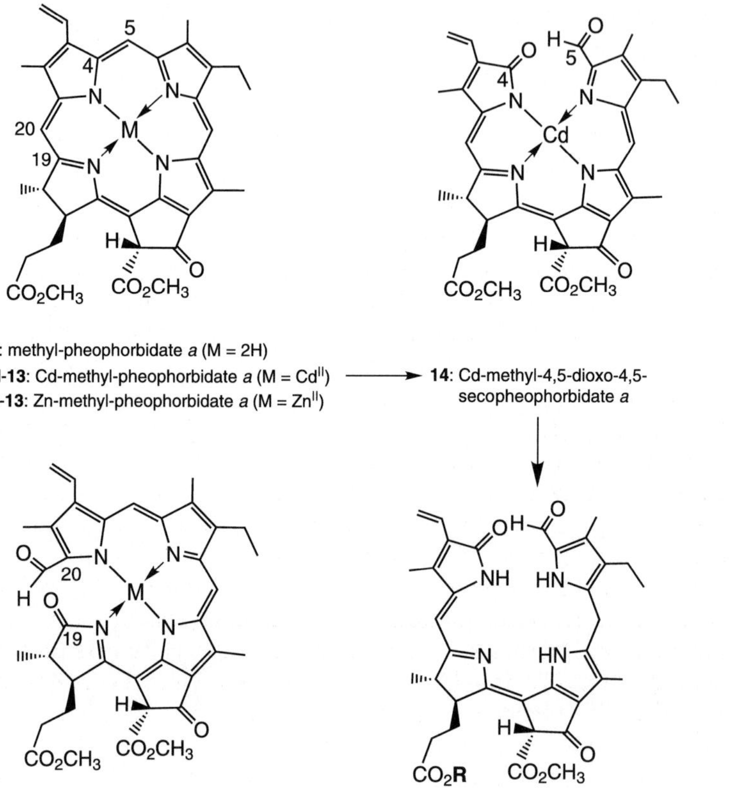

12: methyl-pheophorbidate *a* (M = 2H)
Cd-13: Cd-methyl-pheophorbidate *a* (M = Cd^{II})
Zn-13: Zn-methyl-pheophorbidate *a* (M = Zn^{II})

14: Cd-methyl-4,5-dioxo-4,5-secopheophorbidate *a*

Cd-15: Cd-methyl-19,20-dioxo-19,20-seco-pheophorbidate *a* (M = Cd^{II})
Zn-15: Zn-methyl-19,20-dioxo-19,20-seco-pheophorbidate *a* (M = Zn^{II})

16: methyl-4,5-dioxo-4,5-secopheophorbidate *a* (RCC methyl ester, R = CH$_3$)
11: red chlorophyll catabolite (RCC, R = H)

Scheme 7. Partial synthesis of the elusive red chlorophyll catabolite (RCC, **11**) from pheophorbide *a* (**5a**). Photo-oxygenolysis of Cd-methyl-pheophorbidate *a* (Cd-**13**) gave Cd-methyl-4,5-dioxo-4,5-secopheophorbidate *a* (**14**) (besides a trace of the isomeric Cd-methyl-19,20-dioxo-19,20-seco-pheophorbidate, Cd-**15**); reduction of **14** with sodium borohydride and metal extrusion with dilute aqueous acid provided methyl-4,5-dioxo-4,5-seco-pheophorbidate **16** in good yield; partial hydrolysis of the red diester **16** with pig liver esterase was regio-selective and produced red chlorophyll catabolite **11** (RCC)

References, pp. 37–43

respectively, and catalyzed by the "RCC-reductase", would generate the "primary" fluorescent chlorophyll catabolite (**10**) from the red tetrapyrrole **11** (*62*) (see Scheme 6).

The elusive tetrapyrrole **11** appeared attractive as an intermediate, as it had also the same chromophore structure as some of the red bilinones, which were found to be excreted as final degradation products of the chlorophylls in the green alga *Chlorella protothecoides* (*57, 58, 59*). Based on earlier work for the chemical preparation of red tetrapyrrolic isolate **8** from the green alga *C. protothecoides* via a photo-oxygenolytic opening of the macrocycle of the Cd-methyl pyropheophorbidate (see Scheme 16, Section 3) (*73, 74, 75*), the red tetrapyrrole **10** could be prepared by partial degradation of methyl-pheophorbidate *a* (**12**, the methyl ester of Pheo *a* (**5a**)) in a sequence of five chemical steps (see Scheme 7) (*76*): Photo-oxygenolysis of the Cd-methyl-pheophorbidate *a* (Cd-**13**) gave the Cd-methyl-4,5-dioxo-4,5-secopheophorbidate *a* **14** in approximately 35% yield (besides about 10% yield of the isomeric Cd-methyl-19,20-dioxo-19,20-seco-pheophorbidate (Cd-**15**), see Scheme 2 for atom numbering).

As observed earlier (*74*), under comparable experimental conditions, photo-oxygenolysis of Zn-methyl-pheophorbidate *a* (Zn-**13**) generated the Zn-methyl-19,20-dioxo-19,20-seco-pheophorbidate *a* (Zn-**15**) as the main product (25% yield, see Scheme 7) (*76*). This latter cleavage pattern, with the main cleavage site at the δ-*meso*-position (*i.e.* next to the partially reduced ring D) of the chlorin macro-ring, has been generally observed in photo-oxygenation reactions with chlorins (*8, 19, 75*) (see Section 4 for an interesting and very recent further study (*77*) to this subject). The photo-oxygenation of Zn-**13** led to Zn-**15** as main product, indicative of cleavage between the C(19) and C(20) centers. In contrast, the photo-oxygenolysis of 20-methyl-pheophorbides, such as of bacteriochlorophyll *c*, provided 1,20-dioxo-1,20-secopheophorbidates, indicating cleavage to occur at the C(20) and C(1) carbons (*8, 19, 77*). All of these results agreed with the known preferred reactivity of chlorins with electrophiles at the δ-*meso*-position (*8, 18*).

The cadmium 4,5-dioxo-4,5-seco-pheophorbidate **14** was reduced with sodium borohydride and demetallated with dilute aqueous acid to provide methyl-4,5-dioxo-4,5-secopheophorbidate **16** in about 72% yield. The UV/Vis-spectrum of the weakly fluorescing red diester **16** has prominent absorbance maxima near 500 and 316 nm (*1*). The diester **16** was spectroscopically identified (*76*) with the methylation product (compound **32a**, see Scheme 16) of the diacid **9** from the green alga *C. protothecoides* (*78*). Regioselective, partial hydrolysis of the diester **16** with pig liver esterase occurred practically exclusively at the propionic acid side chain and produced the red chlorophyll catabolite **11** (RCC,

$3^1,3^2$-didehydro-4,5,10,17,18-(22H)-hexahydro-13^2-(methoxycarbonyl)-4,5-dioxo-4,5-seco-phytoporphyrin), a monoacid, in nearly quantitative yield (76).

2.2.3. An Enzyme-bound Red Chlorophyll Catabolite from Enzymatic Oxygenation of Pheophorbide a

With the authentic red tetrapyrrolic RCC (**11**) available as a reference material from the synthetic work (76), an identical red compound was detected in senescent plant material and was identified as an elusive "red" chlorophyll catabolite, when Pheo *a* (**5a**) was incubated with aerated extracts of washed membranes of senescent *Canola* chloroplasts (see Scheme 8) (68, 69). In addition, incubation of chemically pre-

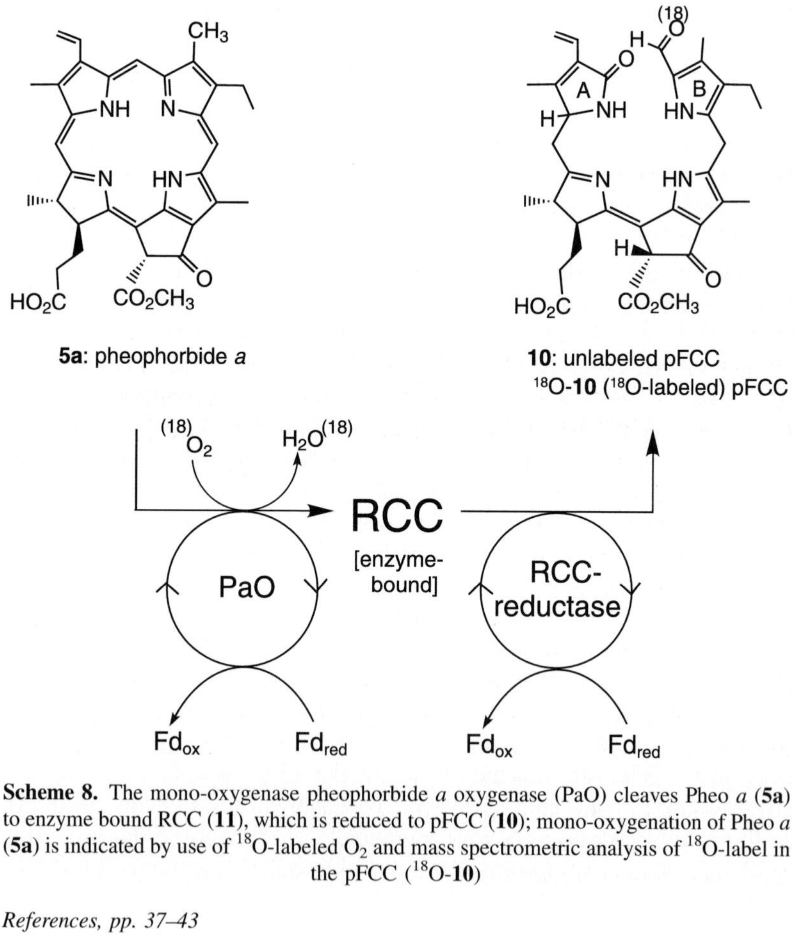

Scheme 8. The mono-oxygenase pheophorbide *a* oxygenase (PaO) cleaves Pheo *a* (**5a**) to enzyme bound RCC (**11**), which is reduced to pFCC (**10**); mono-oxygenation of Pheo *a* (**5a**) is indicated by use of ^{18}O-labeled O_2 and mass spectrometric analysis of ^{18}O-label in the pFCC (^{18}O-**10**)

pared **11** with a preparation of stroma proteins from chloroplasts of senescent cotyledons resulted in the formation of three FCCs, provided that reduced ferredoxin was furnished under anaerobic conditions. These fluorescent compounds had UV/Vis-absorbance properties as the primary fluorescent chlorophyll catabolite **10** (pFCC, $3^1,3^2$-didehydro-1,4,5,10,17,18,20-(22H)-octahydro-13^2-(methoxycarbonyl)-4,5-dioxo-4,5-seco-phytoporphyrin), one of the three fractions displaying also HPLC-characteristics identical to those of **10** (*68, 69*).

The oxygenolytic formation of (enzyme bound) red chlorophyll catabolite **11** from Pheo *a* (**5a**) involved molecular oxygen and was achieved by a single enzyme, an oxygenase termed pheophorbide *a* oxygenase (PaO) (*5, 79*). The activity of PaO, which catalyzes the crucial (and effectively irreversible) cleavage reaction of the porphinoid macrocycle, was low in green leaves and had a considerably higher level in senescent leaves: PaO was thus considered to represent the "key enzyme" of Chl-breakdown (*5, 72*).

An *in vitro* assay helped to characterize the mechanism of PaO: As the oxygenase was known to be inhibited by its tightly binding product **11** (*41*), the analysis was actually carried out with an assay containing both partially purified oxygenase and an extract containing the reductase from oilseed rape (*Brassica napus*, see below), so that the "primary" fluorescent chlorophyll catabolite **10** (pFCC = *Bn*-FCC-2) (*62*) was analyzed as the product of both steps (*72*). In the presence of $^{18}O_2$, the mixture of partially purified enzymes converted Pheo *a* (**5a**) into ^{18}O-labeled pFCC (^{18}O-**10**) containing one ^{18}O-atom per molecule of catabolite, as determined from analysis of the molecular ion by mass spectrometry (see Scheme 8) (*72*). From mass spectral analysis of fragment ions of ^{18}O-**10**, the isotopic label could be localized further to the formyl group at "ring B". As these results indicated the incorporation of one oxygen atom from O_2 at C-5 of the α-*meso* position of **5a**, one of the two oxygen atoms introduced in the oxidation reaction of **5a** to **11** must stem from a different source, most likely (directly or indirectly) from water. Accordingly, PaO was characterized as a monooxygenase (*72*).

PaO is intriguingly specific for Pheo *a* (**5a**) and is located in the chloroplast envelope. It catalyzes the remarkable transformation of **5a** into (a bound form of) RCC (**11**) (*5*). Besides the incorporation of two oxygen atoms, the ring opening at the newly oxygenated sites appears to achieve, all in this step, the formation of two carbonyl functions and the saturation of the "eastern" β-*meso* position. The mechanism of the hypothetical isomerization of the primary enzymatic oxygenation product to the ring-opened (enzyme-bound form of) **11** has not been clarified.

Formally, **11** arises from Pheo *a* (**5a**) by addition of one equivalent each of dioxygen and dihydrogen (see Section 4 for a mechanistic suggestion by Gossauer *et al.* concerning the formation of the related red bilinones (such as **9**) in the green alga *C. protothecoides* (*58, 59, 78*)). The red catabolite **11** inhibits PaO by binding to it in an as yet structurally uncharacterized state. For this reason, significant amounts of RCC (**11**) have never been observed in the course of the senescence processes in fully functional higher plants. Trace amounts of **11** may be found in *in vitro* catabolic experiments, when Chl-breakdown is artificially interrupted (*68, 69*). Alternatively, the absence of the activity of RCC-reductase, the enzyme that catalyzes the reduction of RCC to the pFCC (**10**) or its epimer (*epi*-**10**), in genetically produced deletion mutations of *Arabidopsis thaliana* led to the accumulation of RCC (**11**) or related compounds (*80*), as similarly suspected to occur in plants defective in the "death genes" *acd-1* and *acd-2* (*81, 82*), which are now associated with the reductase (*80*).

2.2.4. Fluorescent Chlorophyll Catabolites from Enzymatic Reduction of the Red Chlorophyll Catabolite

The red chlorophyll catabolite RCC (**11**) is bound strongly to PaO and inhibits it. In an *in vitro* assay, the soluble reductase from oilseed rape converted **11** to the primary fluorescent chlorophyll catabolite pFCC (**10**, $3^1,3^2$-didehydro-1,4,5,10,17,18,20-(22H)-octahydro-13^2-(methoxycarbonyl)-4,5-dioxo-4,5-seco-phytoporphyrin) (*62, 83*). The reductase, which was named red chlorophyll catabolite reductase (RCC-reductase) (*68, 80, 83*), introduced the chiral center C(1) *via* a stereo-selective reduction step. However, early studies with oilseed rape and sweet pepper indicated a remarkable stereo-dichotomy of the respective reductases (see above) (*67, 68, 69*). Screening of a variety of plant species for their type of "primary" FCC revealed the broad existence of two classes of the "RCC-reductases", whose stereo-selectivity was species specific (*84*). At present, the (absolute or relative) configuration at C(1) in the two pFCCs (**10** and *epi*-**10**) is not yet established (*2*). Indeed, the existence of the two epimeric pFCCs (**10** and *epi*-**10**) (see Scheme 6) indicated the absolute configuration at the newly generated chiral center to have no apparent functional relevance (*67, 68, 69*).

The central steps of chlorophyll breakdown in higher plants, which result in the cleavage of the Chl-macrocycle, thus depend on the intimate cooperation of the membrane bound PaO and RCC-reductase: these two effectively coupled enzymatic steps possibly provide an example of "metabolic channeling" (*4, 5, 60, 85*).

References, pp. 37–43

2.2.5. Model Experiments for the Reduction of the Red Chlorophyll Catabolite to Fluorescent Chlorophyll Catabolites

RCC-reductase depends on reduced ferredoxin as electron donor, while (other) cofactors appear not to be involved in its task of reducing enzyme-bound RCC (**11**) to **10** (*83*). At first sight, this observation appeared very puzzling. However, it suggested the possibility, that the bound red catabolite **11** might be sufficiently redox-active as substrate of this reductase, to undergo a ferredoxin-driven reduction to **10** without the help of a reducing cofactor. To test this assumption, the reduction of the methyl ester of the red chlorophyll catabolite ("RCC methyl ester" (**16**) available from partial synthesis (*76*)) was studied in analytical as

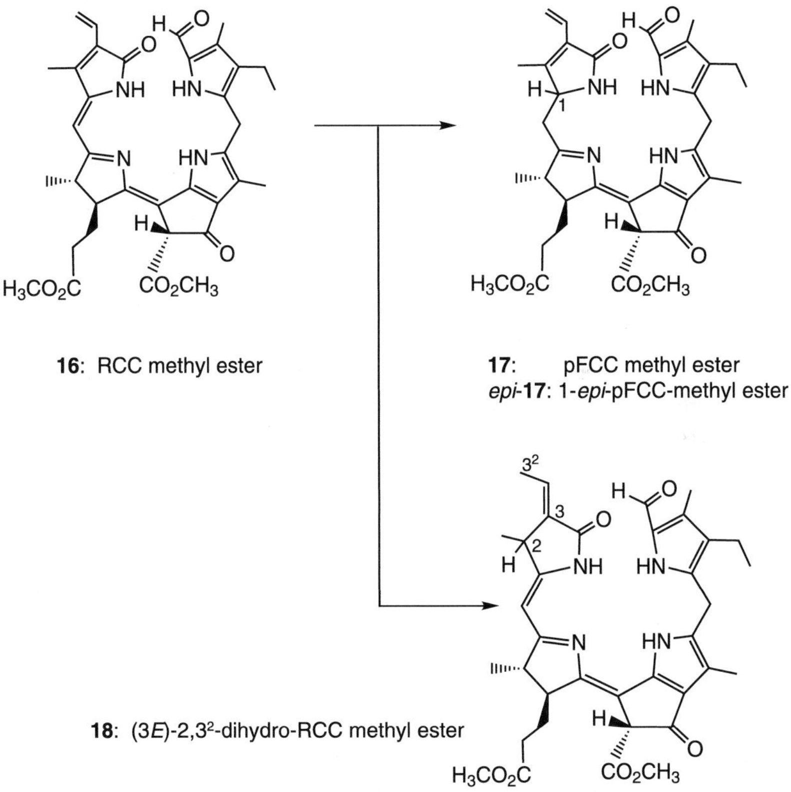

16: RCC methyl ester

17: pFCC methyl ester
epi-**17**: 1-*epi*-pFCC-methyl ester

18: (3*E*)-2,3^2-dihydro-RCC methyl ester

Scheme 9. Electrochemical reduction of RCC methyl ester (**16**) to the methyl esters of pFCC and *epi*-pFCC (**17** and *epi*-**17**), as well as to the regio-isomeric reduction product **18** (and its stereo-isomers)

well as in preparative electrochemical experiments (*86*): Indeed, electrochemical reduction of **16** in methanol and at room temperature reduced about 25% of the starting material into two major (and two minor) compounds displaying the UV/Vis-absorbance properties of pFCC (**10**). The electrochemical reduction proceeded rather stereo-unselectively and provided about 12% each of the strongly luminescent tetrapyrroles **17** and *epi*-**17**, the methyl esters of the two epimeric pFCCs (**10** and *epi*-**10**, see Scheme 9). In addition, about 30% of new reduction products were formed, with a different chromophore structure and a UV/Vis-spectrum showing absorbance maxima near 310 and 420 nm (*86*). Mass spectrometric investigations showed the four main fractions to have the same molecular formula as **17**. The practically non-fluorescent tetrapyrrole **18** and its three stereoisomers were structurally characterized further by NMR spectroscopy. They were found to differ from each other by the stereochemistry at C(2)- and C(13^2) and to be tetrapyrrolic reduction products with an ethylidene functionality at ring A, *i.e.* to be regioisomers of **17** and *epi*-**17** (see Scheme 9). The spectroscopically derived functionalities of the methyl-$3^1,3^2$-didehydro-1,4,5,10,17,18,20,22-octahydro-13^2-(methoxycarbonyl)-4,5-dioxo-4,5-seco-(*22H*)-phytoporphyrin (**17**) and of the methyl-3^1-dehydro-2,4,5,10,17,18,22-heptahydro-13^2-(methoxycarbonyl)-4,5-dioxo-4,5-seco-(*22H*)-phytoporphyrin (**18**, and of their stereo-isomers) are remarkably reminiscent (*86*) of the structures of some phycobilins (*87*), enzymatic reduction products of biliverdin (**19**), such as phytochromobilin (**20**) and 15,16-dihydrobiliverdin (**21**) (*88, 89*). Indeed, RCC-reductases (*5, 83*) show considerable homology with ferredoxin-dependent biliverdin reductases (*89, 90*).

The electrochemical model experiments, therefore, support the idea, that RCC (**11**) might be inherently sufficiently redox-active to undergo ferredoxin-driven and enzyme-mediated reduction to **10** or *epi*-**10** (*86*). The reduction of RCC by RCC-reductase thus may come about in single electron reduction and protonation steps. If so, RCC-reductase would have the role (i) of docking both enzyme-partners, product loaded pheophorbide oxygenase (*i.e.* with bound **11**) and reduced ferredoxin, (ii) of mediating the electron transfer reactions, and (iii) of controlling properly the regio- and stereo-selective protonation (at C(20) and C(1)) of the protein bound tetrapyrrolic reduction intermediates. In this model, the reductase as such would not carry out the reduction steps; it would, however, help directing them in an optimal way and play the part of a "chaperone" in a redox reaction (*86*). On the other hand, the homology of RCC-reductase and of some biliverdin reductases (*89, 90*), their related demand for ferredoxin, and the relationships of the biochemical transformations catalyzed by these enzymes are all rather striking: they

References, pp. 37–43

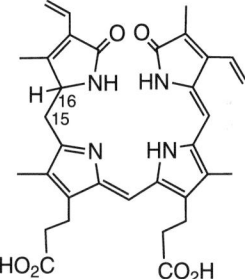

19: biliverdin   **39**: bilirubin

20: (3Z)-phytochromobilin   **21**: 15,16-dihydrobiliverdin

Scheme 10. Biliverdin (**19**), bilirubin (**39**) and isomeric, natural dihydro-biliverdins, phytochromobilin (**20**) and 15,16-dihydro-biliverdin (**21**, bilane type atom numbering, see (87, 88))

point at an organizational similarity in higher plants of heme-breakdown (*via* biliverdin (**19**)) towards the phycobilins (such as phytochromobilin (**20**) or 15,16-dihydro-biliverdin (**21**)) and Chl-breakdown (*via* RCC (**11**) and pFCC (**10**)) (see Scheme 10) (*86, 89*).

2.2.6. Non-fluorescent Colourless Chlorophyll Catabolites

The constitution of *Hv*-NCC-1 (**2**, $3^1,3^2,8^2$-trihydroxy-1,4,5,10,15,20-(22*H*,24*H*)-octahydro-13^2-(methoxycarbonyl)-4,5-dioxo-4,5-seco-phytoporphyrinate (see Scheme 2) gave first clues on the basic transformations involving the Chl-chromophore (*1, 2, 4, 10*). When, in addition, the structure of the fluorescent chlorophyll catabolite pFCC (**10**) was revealed, an isomerization of the chromophore of the FCCs into that of the corresponding colourless and non-fluorescent chlorophyll catabolites (NCCs) was suggested to be a likely "final" transformation (*1*, *62*). The characteristic complete de-conjugation of the four pyrrolic units of the tetrapyrrolic NCCs could (possibly) result from non-enzymatic tautomerization reactions involving the chromophoric system of rings C and D of the FCCs, the final steps in the complex transformation of the chromophoric system of the highly coloured Chls into that of the colourless NCCs (*56*).

2.2.7. A Non-enzymatic Tautomerization Achieves the "Final" Transformation of Fluorescent Chlorophyll Catabolites to Non-fluorescent Colourless Chlorophyll Catabolites

The fluorescent chlorophyll catabolites, such as pFCC (**10**), were observed not to accumulate during chlorophyll breakdown in senescent leaves (*24*). The indicated further transformation of the FCC chromophore to those of non-fluorescent chlorophyll catabolites (NCCs) was suggested to possibly be the result of a non-enzymic isomerization (*56*, *62*). In analogy to the results of studies on the tautomerization chemistry of a range of hydro-porphinoids (*91*), the isomerization of the chromophore of FCCs into that of NCCs was judged to be rather favorable, thermodynamically. The complete de-conjugation of the four pyrrolic units, characteristic of the tetrapyrrolic NCCs, thus may occur in the course of natural chlorophyll breakdown under rather mild and, possibly, even without catalysis by (an) enzyme(s) (*56*).

Indeed, the generation of the primary FCCs (**10** and *epi*-**10**) in the chloroplast, and the spatial localization of the NCCs to the vacuoles (*24*), both suggested a transport in the senescent leaf cell during chlorophyll breakdown and the site of the hypothetical FCC to NCC isomerization to possibly coincide with the vacuolar system. The acidic medium in these organelles could also provide the required weakly acidic medium for a hypothetical non-enzymatic conversion of an FCC into the corresponding NCC (*4*). Considering the functional groups present in the typical NCCs (such as *Hv*-NCC-1, **2**), further peripheral modifications of the pFCCs by enzymes within the chloroplast were taken into account. However, considering the variability of the structures of the known NCCs, the hypothetical FCC- to NCC-isomerization (which cuts into two parts and de-conjugates the main chromophore of the FCCs) may occur before, in parallel or after such further modification reactions. The export of functionalized FCCs from the chloroplast and their carrier mediated entry in the vacuoles were considered to be supported by the availability of polar peripheral groups (*3*, *4*). The recent observation of

References, pp. 37–43

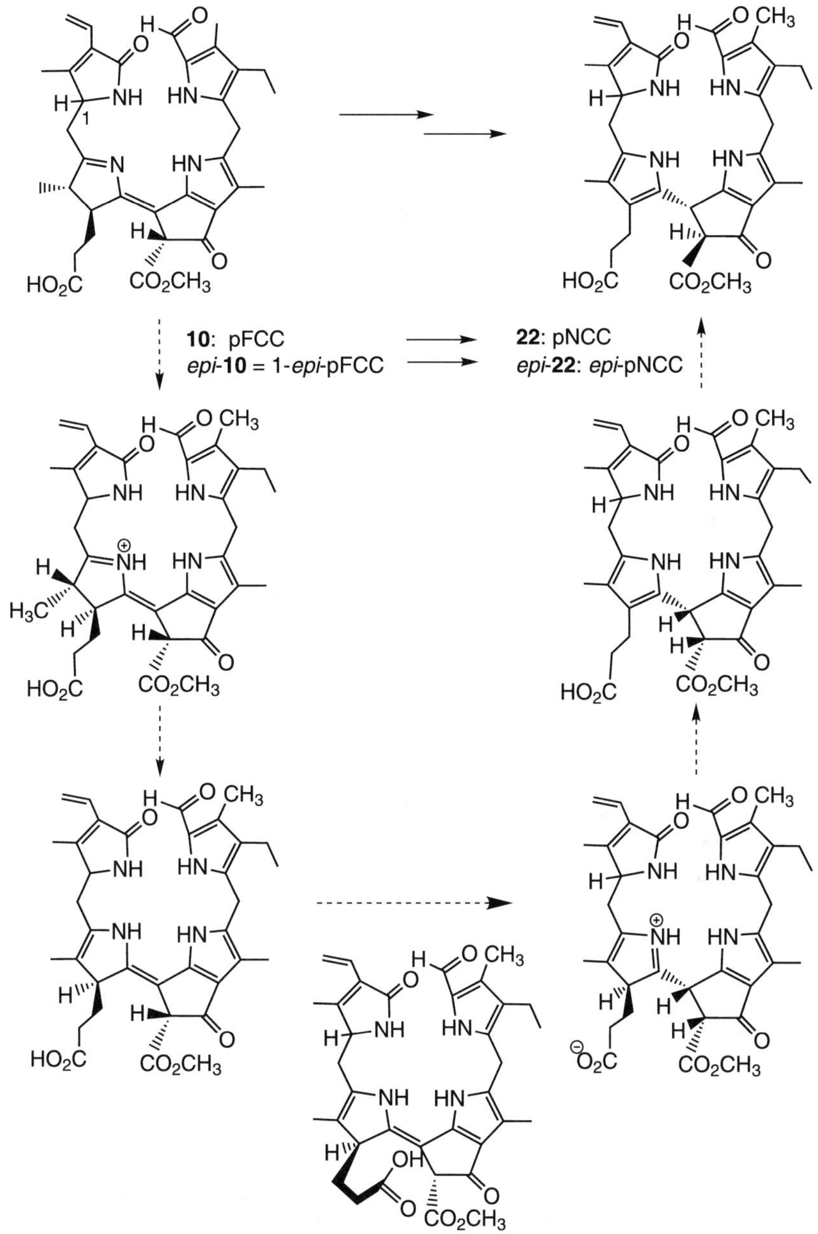

Scheme 11. Non-enzymatic isomerization of *epi*-pFCC (*Ca*-FCC-2, *epi*-**10**) to the "primary" NCC *Cj*-NCC-2 (*epi*-**22**) see (*56*) and of the pFCC (**10**) to the NCC (**22**) and stereochemical assignment in natural NCCs, derived from the suggested isomerization mechanism *via* an intramolecular protonation at the *re*-face of C15 (with a proton mediated *via* the propionic acid side chain at C(17), see proposed reactive conformation in the lower formula) (*56*)

more polar compounds displaying fluorescence properties as those of the pFCCs in *Arabidopsis thaliana* would also support the view (*80*) that the vacuoles, the final storage vessel for the NCCs, would be the likely sites for the final isomerization of FCCs to NCCs. Indeed, chemical experiments with the pFCC *epi-***10**, available from the *in-vitro* transformation system from senescent *Capsicum annuum* (*67*), showed a considerable readiness of this pFCC to undergo acid-induced, stereo-selective tautomerization to the corresponding NCC *epi-***22** in the absence of enzymes (see Scheme 11) (*56*).

The NCC *epi-***22** turned out to be identical with a non-polar NCC from senescent leaves of the tree *Cercidiphyllum japonicum* and named *Cj*-NCC-2, a $3^1,3^2$-didehydro-1,4,5,10,15,20-(22H,24H)-octahydro-13^2-(methoxycarbonyl)-4,5-dioxo-4,5-seco-phytoporphyrinate and an isomer the of the pFCC (*epi-***10**, see Scheme 11) (*56*). The NCC *epi-***22** lacked the characteristic oxygen atom attached at carbon 8^2, at the ethyl side chain of ring B (see Scheme 11). As an isomerization of the pFCC *epi-***10** directly gave the NCC *epi-***22**, it was considered a "primary" NCC (or pNCC) of *Cercidiphyllum japonicum* (*56*). The tendency of pFCC (**10**) to tautomerize under mild conditions was also investigated in recent further studies. Both of the primary FCCs turned out to undergo readily the stereo-selective, acid-catalyzed isomerization to the corresponding NCCs, in contrast to the dimethyl ester **17** and *epi-***17** (indication of participation of the propionic acid function, see Scheme 11) (*92*).

2.2.8. Peripheral Functional Groups and Conjugations Found in Non-fluorescent Colourless Chlorophyll Catabolites

The structures of most natural NCCs, such as of *Hv*-NCC-1 (**2**) or of *Cj*-NCC-1 (**23**), indicate further refunctionalization reactions, most of which are likely to be enzyme-catalyzed. A remarkable peripheral hydroxylation at the terminal position of the ethyl side chain at ring B is systematically indicated by the published structures of NCCs (such as, e.g. *Hv*-NCC-1, **2**) (*1*, *2*, *10*). This peripheral hydroxylation, for which an enzyme-catalyzed reaction appears to be required, may serve the purpose of increasing the polarity of the catabolites and of providing an anchor point for further, secondary refunctionalization with hydrophilic groups (*4*). A uniform picture concerning the timing and the spatial localization in the leaf cell of the enzymatic activities for hydroxylation of the ethyl group at carbon 8 and for oxidation (with di-hydroxylation) of the vinyl side chain at carbon 3 is not yet apparent. Possibly, even the discrimination between FCCs or NCCs as enzyme substrates by some of these enzymes may not be high (5). However, the mentioned localization of the NCCs in the vacuoles of senescent plant leaves is consistent with the requirement for intriguing transport mechanisms.

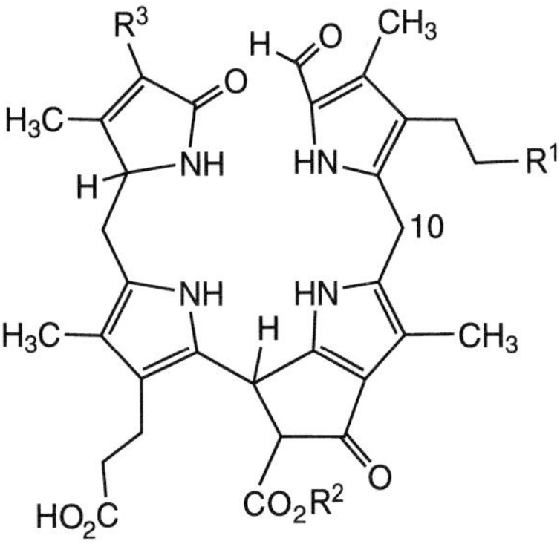

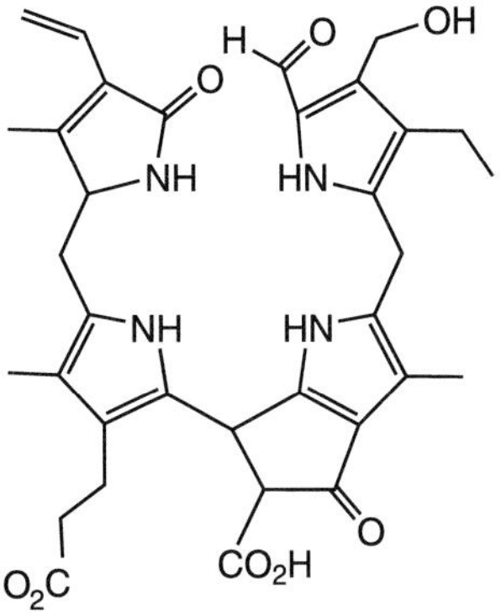

28c: *At*-NCC-3

Compound		R¹	R²	R³
2	*Hv*-NCC-1	OH	CH₃	CH(OH)CH₂OH
22	*Cj*-NCC-2	H	CH₃	CH=CH₂
23	*Cj*-NCC-1	OH	CH₃	CH=CH₂
24a	*Bn*-NCC-1	O-Mal	H	CH=CH₂
24b	*Bn*-NCC-2	O-β-Glc	H	CH=CH₂
24c	*Bn*-NCC-3	OH	H	CH=CH₂
24d	*Bn*-NCC-4	H	H	CH=CH₂
25a	*So*-NCC-1	OH	H	CH(OH)CH₂OH
25b	*So*-NCC-2	OH	CH₃	CH(OH)CH₂OH
25c	*So*-NCC-3	OH	H	CH=CH₂
25d	*So*-NCC-4	OH	CH₃	CH=CH₂
25e	*So*-NCC-5	H	CH₃	CH=CH₂
26a	*Nr*-NCC-1	O-β-(6′-O-Mal)Glc	CH₃	CH=CH₂
26b	*Nr*-NCC-2	O-β-Glc	CH₃	CH=CH₂
27a	*Zm*-NCC-1	O-β-Glc	CH₃	CH(OH)CH₂OH
27b	*Zm*-NCC-2	O-β-Glc	CH₃	CH=CH₂
28a	*At*-NCC-1	O-β-Glc	H	CH=CH₂
28b	*At*-NCC-2	OH	H	CH=CH₂
28d	*At*-NCC-4	O-β-Glc	CH₃	CH=CH₂
28e	*At*-NCC-5	H	H	CH=CH₂

Abbreviations: Mal = malonyl; Glc = glucopyranosyl

Scheme 12. Constitution of non-fluorescent chlorophyll catabolites (NCCs) from higher plants (*1*)

The catabolite *Hv*-NCC-1 (**2**) was obtained from de-greened primary leaves of the monocot barley (*Hordeum vulgare*), which were forced to senesce in permanent darkness (*10, 25, 27*). In naturally de-greened senescent cotyledons of the dicot canola (*Brassica napus*), NCCs (*Bn*-NCCs) also were found. This was of particular interest, as the senescence of these cotyledons occurred under natural growth conditions (*93, 94*). Four NCCs were found in the cotyledons of oilseed rape, termed *Bn*-NCCs (*Bn*-NCC-1 (**24a**), *Bn*-NCC-2 (**24b**), *Bn*-NCC-3 (**24c**) (*48, 95*), and the less polar *Bn*-NCC-4 (**24d**), as recently identified by mass spectrometry (*96*)). Most notably the common basic structure of the three (more polar) *Bn*-NCCs (**24a–24c**) were revealed through spectroscopic investigations to be the same as the one of *Hv*-NCC-1 (**2**) from barley (see Scheme 12) (*48, 95*). The three *Bn*-NCCs differed from the catabolite **2** of barley merely by some peripheral (re)functionalizations. *Bn*-NCC-3 (**24c**), might be the biosynthetic precursor of the more polar analogues (**24a, 24b**) (*48, 94, 95*): The observed primary alcohol function at position 8^2 of *Bn*-NCC-3 (**24c**) appeared to represent an anchor point for further secondary conjugations with hydrophilic moieties, such as with a malonyl group in *Bn*-NCC-1 (**24a**) and with a β-glucopyranosyl group in *Bn*-NCC-2 (**24b**). The esterification of NCCs with a free 8^2-hydroxyl function with malonic acid has been achieved with a protein preparation from *Canola* cotyledons and malonyl-CoA as substrate (*97*). The *Bn*-NCCs accounted for practically all of the Chls broken down in the senescent cotyledons of oilseed rape.

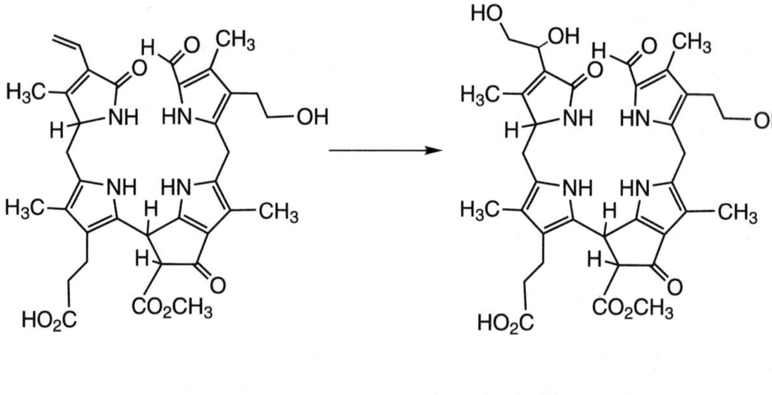

23: *Cj*-NCC-1

25b: *So*-NCC-2

Scheme 13. Stereo-unselective chemical dihydroxylation of *Cj*-NCC-1 (**23**) gives *So*-NCC-2 (**25b** and its C(3^2)-epimer), which is also the C(1)-epimer of *Hv*-NCC-1 (**2**)

Non-fluorescent chlorophyll catabolites (NCCs) were found in a variety of senescent higher plants, such as the autumn leaves of sweet gum (*Liquidambar styraciflua*, see Scheme 12) (*49*) and of the tree *Cercidiphyllum japonicum* (*Cj*-NCCs, see Schemes 11–13) (*50, 56*), in naturally de-greened leaves of spinach (*So*-NCCs **25a–25e**, see Scheme 12) (*51, 52*), of tobacco (*Nr*-NCCs **26a, 26b**) (*53*), of corn (*Zm*-NCCs **27a, 27b**) (*54*), *etc*. All NCCs isolated, so far, from a variety of de-greened plants represent linear tetrapyrroles of uniform basic build-up (see Schemes 2 and 12) and relate to Chl *a* (**1a**) rather than to Chl *b* (**1b**) (*1, 2, 3*). However, among the five NCCs from artificially de-greened leaves of *Arabidopsis thaliana* (the *At*-NCCs **28a–28e**) (*36, 60*), an NCC of intermediate polarity (*At*-NCC-3, **28c**) carried a hydroxyl-methyl group at position 7 and an unmodified ethyl side chain at carbon 8 (*36*) (see Scheme 12). The mechanistic explanation for this remarkable exception from the observed hydroxylation pattern is still lacking (*36*).

So-NCC-2 (**25b**), the most abundant of the five NCCs detected in spinach, had the same constitution as the catabolite from barley, *Hv*-NCC-1 (**2**) (*51*). Both of these isomeric NCCs can result (in a formal sense) from an enzymatic dihydroxylation at the vinyl group at ring A. With osmium tetroxide, the catabolite *Cj*-NCC-1 (**23**) (or its methyl ester **29**) was stereo-unselectively dihydroxylated at the corresponding vinyl group. One of the dihydroxylation products of **23** proved to be identical with *So*-NCC-2 (**25b**), whose configuration at C(1) thus differed from that of *Hv*-NCC-1 (**2**) (see Scheme 13) (*51*).

A common feature of the *Bn*-NCCs and of several other NCCs (see Scheme 12) is the presence of a free β-ketocarboxylic acid group at $C(13^2)$ of the characteristic cyclopentanone moiety (*48, 94, 95*). In contrast, the 13^2-methyl ester function of the Chls is still present in a group of other NCCs, such as *Hv*-NCC-1 (**2**) (see Scheme 12) (*1, 2*). For most given plant species, the 13^2-methyl ester function was found in all its NCCs (see *e.g. Cj*-NCCs and *Nr*-NCCs) (*53, 56*) or it was absent (see *e.g.* the *Bn*-NCCs **24a–24c**) (*48, 94, 95*). In contrast, the substitution pattern at $C(13^2)$ was non-uniform in naturally de-greened leaves of spinach: *So*-NCC-2 (**25b**) and *So*-NCC-3 (**25c**) carry a methyl ester function, *So*-NCC-4 (**25d**) a free carboxylic acid group at position $C(13^2)$ (see Scheme 12) (*51, 52*). As the pFCC **10** was observed in de-greened cotyledons of oilseed rape, enzymatic hydrolysis of the 13^2-methoxycarbonyl group in the course of the formation of the *Bn*-NCCs (**24a–24c**) in this plant is indicated to occur at the stage of the FCCs or later. Treatment of the pFCC **10** by an active extract of soluble enzymes from de-greened cotyledons of oilseed rape produced an FCC with significantly higher polarity, to which the structure of the $3^1,3^2$-didehydro-1,4,5,10,17,18,20-(22H)-octahydro-13^2-(carboxy)-4,5-dioxo-4,5-seco-phytoporphyrin (**30**, a 13^2-demethyl-pFCC, see Scheme 14) was tentatively assigned, based on mass spectrometric data (*2*, *94*). The same extract from senescent cotyledons of oilseed rape did not hydrolyze the methyl ester function in several NCCs, indicating hydrolysis of the 13^2-methoxycarbonyl function in these senescent leaves to occur at the stage of the FCCs (*2*, *94*). In artificially de-greened leaves of *A. thaliana* three FCCs were similarly identified tentatively, which were more polar than the pFCC **10** and were thus also indicated to carry bipolar functional groups (*60*). All in all, the situation concerning the timing of the corresponding enzyme-catalyzed modifications is not yet clear and may differ from one plant species to the other. Indeed, in naturally de-greened leaves of spinach the simultaneous appearance of methyl ester and of free acid forms of the C13^2 β-ketocarboxylic acid grouping in the *So*-NCCs **25a–25e** also suggests the hydrolysis of the corresponding methyl ester function to occur at a rather late stage (*48*, *94*, *95*). These findings indicate modified Pheo *a* derivatives not to be involved in Chl-breakdown in these higher plants (*2*), such as the ones observed in *Chenopodium album* (*i.e.* pyropheophorbide *a* (Pyropheo *a*, **6**) and 13^2-carboxy-pyropheophorbide *a* (**7**)) (*43*).

The hydroxylation of the terminal position of the ethyl group on ring B is a most remarkable modification among the polar groups "introduced" in NCCs. As noted above, the observed primary alcohol function represents a suitable function for further secondary conjugations with hydrophilic moieties (see Scheme 13), which possibly are required for the purpose of intra-organellar transport to the vacuoles (*3*, *98*). Esterification and glucosylation (as first seen in **24a** and **24b**) (*48*, *94*, *95*) are reminiscent of many secondary plant metabolites (*99*) which are, like NCCs, deposited in the vacuoles (*3*, *98*, *100*).

2.2.9. Evidence for Further Breakdown of the Non-fluorescent Colourless Chlorophyll Catabolites in Higher Plants

Endogenous breakdown of chlorophyll in senescent plant produces NCCs as the apparent "final" stage of a rapid "detoxification" process (*3*, *5*, *85*). In senescent leaves of higher plants NCCs accumulate in the vacuoles (*98*, *100*) and in various de-greened leaves, the amount of NCCs corresponded roughly to the calculated amount of Chls (*a* and *b*) present initially in the green leaf (*e.g.* the *Bn*-NCCs in the cotyledons from oilseed rape (*48*) the *Pc*-NCCs in de-greened leaves of the pear tree (*55*)). Likewise, in senescent leaves of barley and of French beans (*Phaseolus vulgaris* L.), the total content of NCCs appeared not to decrease strongly over a time of several days (*25*, *93*).

References, pp. 37–43

It is unclear, at present, whether NCCs, the colourless tetrapyrrolic remnants of the Chls in the senescent leaves, have a further function in the plant. Indeed, NCCs were recently also identified in fruit (in peels of pears and apples) (*55*). In addition, NCCs were recognized to be rather effective antioxidants (*55*). Both findings are suggestive of a further possible physiological role in the ripened fruit (where their amounts do not come up for the Chls present initially in the green fruit) or in the senescent leaf (*55*). Evidence of tetrapyrrolic products of further degradation of NCCs was provided by the identification of colourless urobilinogenoidic linear tetrapyrroles, described as the two stereoisomers **31** and *epi*-**31** (see Scheme 15) (*101*) in extracts of de-greened primary

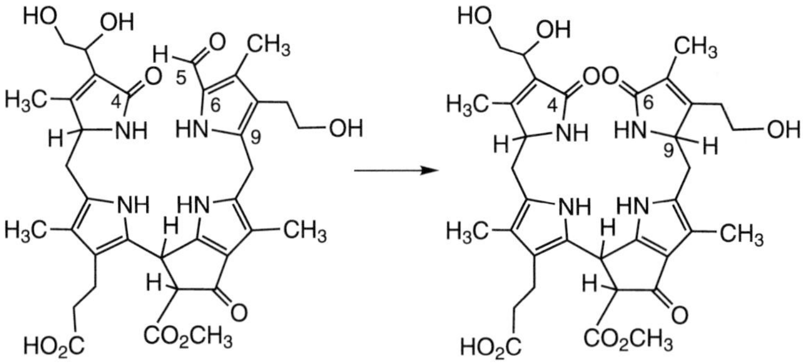

2: *Hv*-NCC-1

31
epi-**31**: C(9)-epimer of **31**

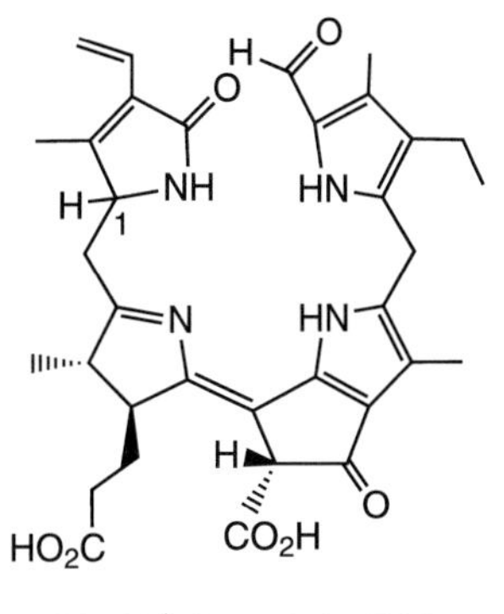

30: 13^4-demethyl-pFCC

Scheme 14. Constitutional formulae of the polar FCC **30**, of *Hv*-NCC-1 (**2**) and of its oxidative deformylation products **31**, *epi*-**31**

leaves of barley. The tetrapyrroles **31** and *epi*-**31** were associated with further degradation of *Hv*-NCC-1 (**2**), from which their constitution differs on account of the absence of the formyl group derived from the α-*meso* position of Pheo *a* (**5a**) (*101*).

The tetrapyrroles **31** and *epi*-**31** were suggested to arise from further endogenous (yet possibly non-enzymatic) transformation of the NCCs in the tissue of the senescent barley leaves. Oxidative loss of the formyl group from related linear tetrapyrroles has been noted (*101*). The original characterization for *Hv*-NCC-1 (**2**) as a "rusty" pigment also pointed to the readiness of these reduced linear tetrapyrroles to undergo spontaneous reactions, which become manifest by the appearance of the rust colour (*3, 4, 25*). Clearly, these and other transformations, such as the one of *Hv*-NCC-1 (**2**) to the two tetrapyrroles **31**/*epi*-**31**, may reflect further degradation of the NCCs in the senescent tissue.

Further breakdown to mono-pyrrolic oxygenation products as further remains of Chls have also been considered (*3, 102*). These studies received further support from recent work by Shioi and coworkers, who obtained evidence for the presence of hematinic acid (4-methyl-2,5-dioxo-2,5-dihydropyrrole-3-propionic acid), ethyl-methyl-maleimide and a putative bicyclic degradation product of the ring-C-E section of Pheo *a* (*103*).

3. Chlorophyll Catabolites from the Green Alga *Chlorella protothecoides*

The green alga *Chlorella protothecoides* was shown earlier to excrete red pigments when grown in nitrogen-deficient and glucose-rich medium (*104, 105*). These red pigments were subjected to structural studies in the laboratory of Gossauer (reviewed in (*58, 59, 78*)), where they were determined to be linear tetrapyrroles. Interestingly, the deduced structures of the red catabolites from the green alga indicated them to also correlate to the Chls by an oxygenolytic cleavage of the macroring at the "northern" α-*meso*-position. In contrast to the plant systems, the red catabolites were found to be derived from Chl *a* (**1a**), as well as from Chl *b* (**1b**) (see Scheme 15) (*58, 59, 106*). Subsequent investigations indicated that the diacid **9** was the authentic product of enzymatic catabolism in *C. protothecoides* (*58, 59*), rather than monoacids, such as **8a** and **8b**, which were isolated and identified originally (*57*) as the (di)methyl esters **32a** and **32b**. These observations may point to the relevance of the enzymatic hydrolysis of the 13^2 methyl ester functionality of the Pheos **5a**/**5b** in *C. protothecoides*, similar to the situation in *Chenopodium album* (*47*). A non-enzymatic decarboxylation of β-keto acids, such as **9**, may readily occur, and, consequently, the methyl esters **32a** and **32b** are likely to be artefacts of the original isolation procedure (*58, 59*).

Isotopic labeling studies with $^{18}O_2$ and mass spectrometric analysis of the excreted pigment as the ^{18}O-labeled methyl ester **32a**, clearly indicated incorporation of only one ^{18}O-atom (from molecular oxygen) (*73*). From analysis of a fragment, the ^{18}O-label was assigned to the formyl group derived from the meso-carbon of Chl. This result suggested the hypothetical ring cleaving enzyme of the green alga to be a mono-oxygenase (*73*), whose direct substrate(s) and product(s) are not

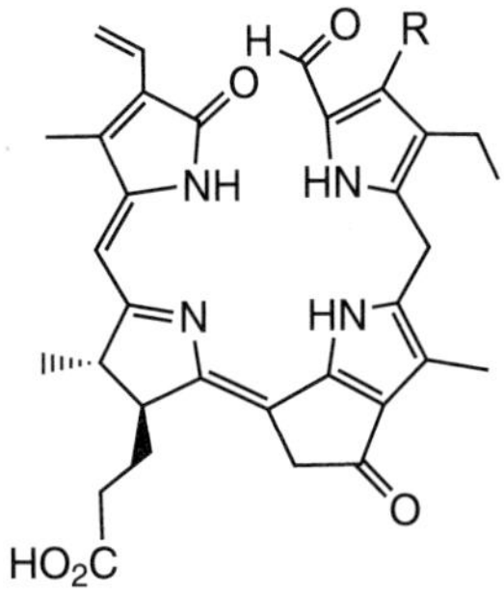

8a: R = CH₃
8b: R = CH=O

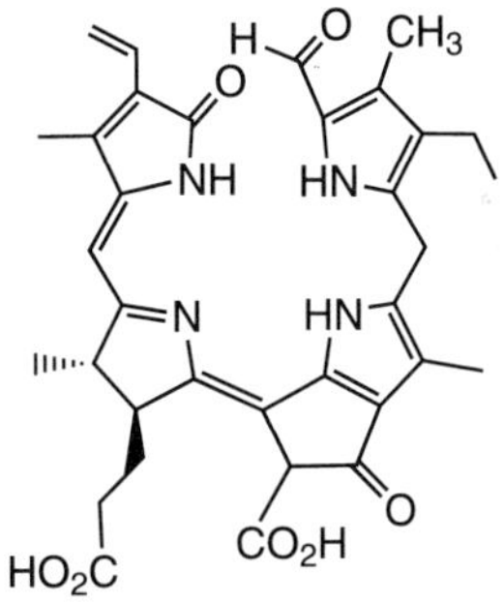

9

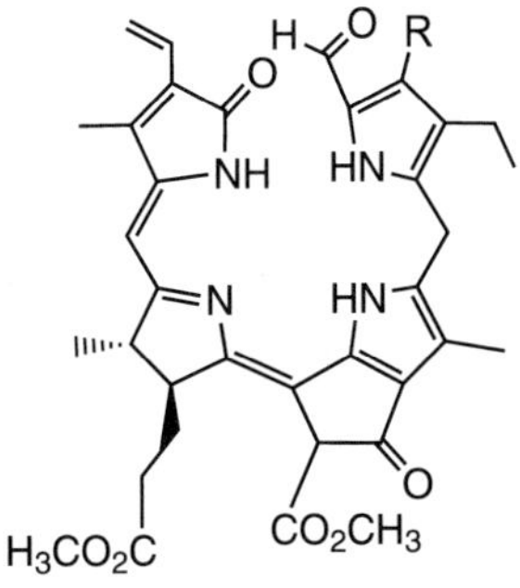

32a: R = CH₃
32b: R = CH=O

Scheme 15. Red tetrapyrrolic degradation products of Chl *a* (**1a**) and Chl *b* (**1b**) from *C. protothecoides*. Isolated monoacids **8a** and **8b** and diacid **9** and derived dimethyl esters (**32a** and **32b**)

known. Further studies concerning the incorporation of deuterium label in the course of the degradation of the Chls in this green alga, showed highly stereo-selective attachment of one hydrogen atom (from water) at the "eastern" β-*meso* position of the red isolate **32a**, indicating that this step in the formation of the red catabolites most likely occurs under control of an enzyme (*107*). The formation of the red Chl-catabolites in the green alga *C. protothecoides* has been suggested to result from hydration of an epoxide intermediate and subsequent rearrangement (*58, 59, 78*). The structural resemblance of the red intermediates from *Chlorella* and the red plant catabolite RCC (**11**), as well as the apparent similarity of the oxygenation mechanisms in chlorophyll breakdown in higher plants (*72*) and in the green alga (*73*) indicate a biochemical relationship. Both of the mono-oxygenases (from higher plants and *C. protothecoides*) may display comparable catalytic properties. Two notable differences concern the substrate specificity and the requirement of a second enzymic reaction (catalyzed by RCC-reductases) in the case of chlorophyll breakdown in higher plants (*2, 72*). The latter enzyme is not known from the green alga, which disposes of its red catabolites by simple excretion, a process which is hardly possible in the case of the vascular plants.

4. Chlorophyll Catabolites from Marine Organisms

Photosynthetic organisms are widely occurring in the oceans (*108, 109*). In contrast to the information now available on chlorophyll catabolism in two green algae and in several higher plants, little is known about the fate of the chlorophylls (or bacteriochlorophylls) from marine organisms. One exception concerns the luciferin of the dinoflagellate *Pyrocystis lunula*, which was suggested earlier to be structurally related to chlorophyll (*110*). The constitution of this colourless, luminescent compound **33a** and of two air oxidation products (**33b** and **33c**) was elucidated with the help of spectroscopic and of chemical degradation methods in the laboratory of Y. Kishi (see Scheme 16) (*21*). Likewise, the bioluminescent transformation of the luciferin **33a** by the dinoflagellate luciferase was shown to lead to the oxidation product **33d**. A related study concerned the structure of the light emitter from krill (*Euphasia pacifica*), which was assigned the structure of the related linear tetrapyrrole **34a** (and which is also readily air oxidized – to **34b**) (*20*). Both luminescent compounds (**33a, 34a**) were thus confirmed to have structural features of Chl derivatives, of 1,20-dioxo-1,20-secopyropheophorbides, in particular. Both these linear tetrapyrroles appear

References, pp. 37–43

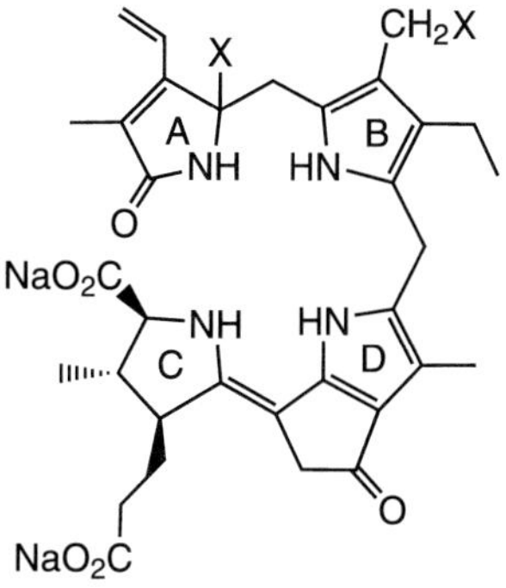

33a: X = H
34a: X = OH

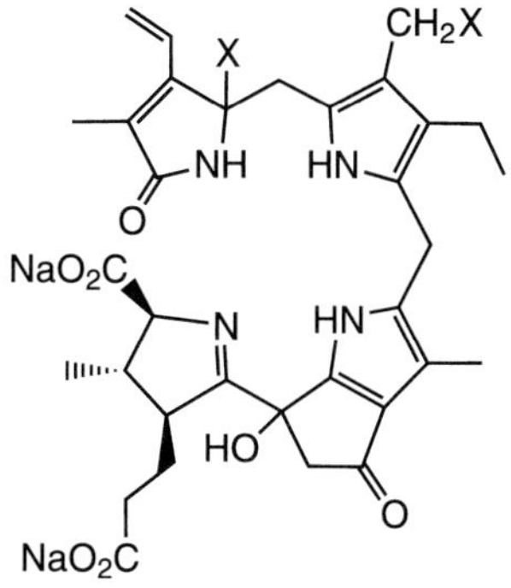

33b: X = H
34b: X = OH

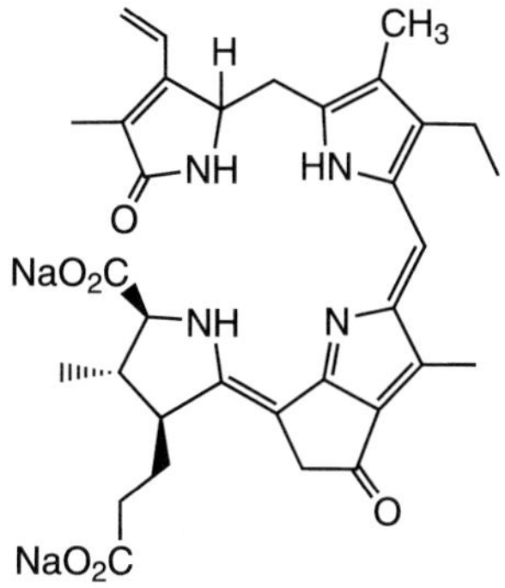

33c

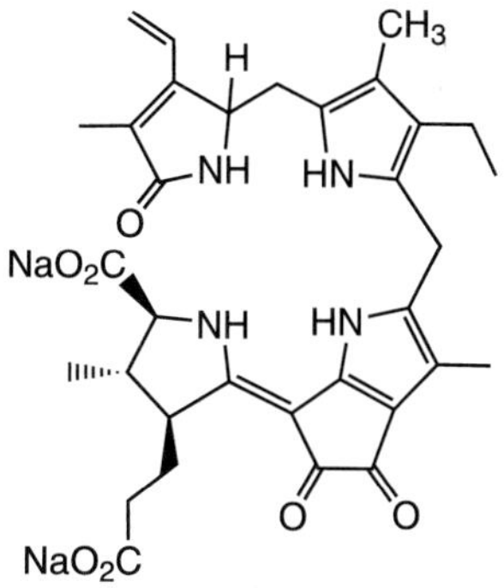

33d

Scheme 16. Formulae of chlorophyll catabolites (**33a**, **34a**) from marine organisms, of their air oxidation products (**33b**, **33c** and **34b**) and of the main product (**33d**) from the luciferin reaction

to arise by an oxygenolytic cleavage at the "western" δ-*meso* position from their natural Chl-precursor(s).

Indeed, recent studies by Kishi and coworkers on the photo-oxygenolysis of the 20-methoxy-pyropheophorbide **35** have confirmed the assumed tendency of such substituted pheophorbides (see *e.g.* (*75*)) to undergo oxygenolytic cleavage of the chlorin macro-ring at the "western" *meso*-position, between C(20) and C(1), and providing synthetic access to the 1,20-seco-pyropheophorbidate **36** (see Scheme 17) (*77*).

As a model for the dipyrrolic chromophore fragment of dinoflagellate luciferin the tri-cyclic pyrrole derivative **37** was prepared by

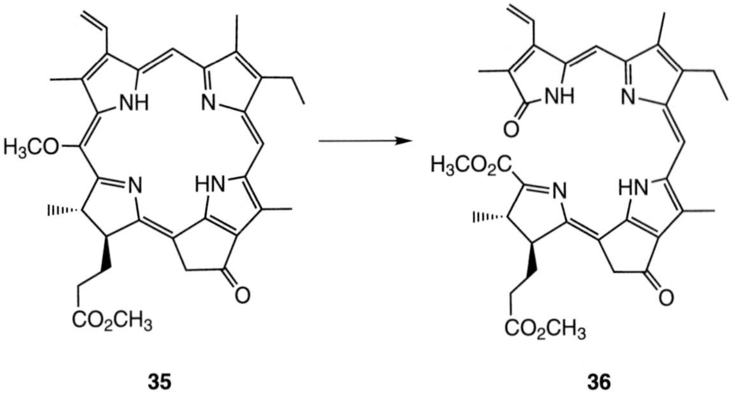

Scheme 17. Photo-oxygenolytic opening of the 20-methoxy-pyropheophorbidate **35** to the 1,20-dioxo-1,20-seco-phytoporphyrinate **36**

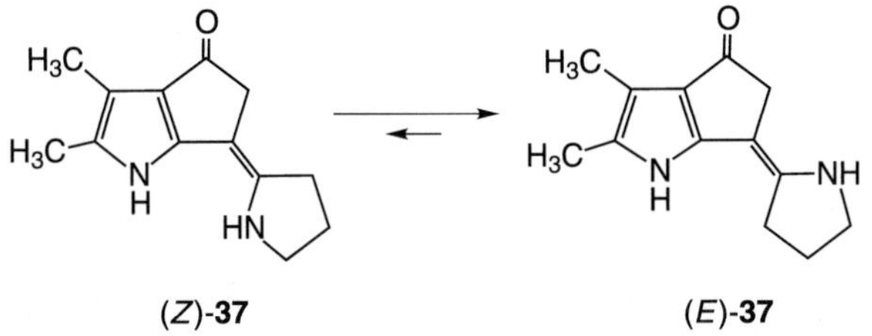

Scheme 18. Tri-cyclic model compounds **37** for the C,D-segment of the tetrapyrroles **33a/34a**

chemical synthesis (see Scheme 18) (*111*). Spectroscopic studies of (*E*)-**37** and (*Z*)-**37** (the (*E*)- and (*Z*)-isomers of **37**) provided firm support for the (*E*)-configuration at the C(15)-C(16) double bond of the natural dinoflagellate luciferin **33a** (*112*).

5. Conclusions and Outlook

In the last fifteen years, Chl catabolism has turned from a major "biological enigma" (*8, 9*) to a thriving research field (*2, 78, 85*). All of the main chemical studies on Chl-breakdown have identified linear tetrapyrroles as the isolated products from (ring-opening) breakdown of the Chls and have concerned investigations with higher plants (*2, 7, 113*), green algae (*58*), and marine organisms (*21*). In spite of the first contribu-

References, pp. 37–43

tions to this last subject (*21*), the fate of (bacterio)chlorophylls available in marine systems is still far from being revealed. In fact, considering the absence of molecular oxygen and the resulting anaerobic environment in deep-sea water, non-oxygenolytic mechanisms may be the dominant form of degradation of chlorophylls from marine photosynthetic organisms. Consistent with such a scenario, the important observation of ubiquitous "geo-porphinoids" in petroleum and shale oil (notably the vanadyl- "deoxo-phylloerythroetioporphyrin" **38a**, a 17^2-decarboxy-13^1-deoxo-phyto-porphyrinate, discovered in the early 20th century) (*114, 115*) may well be relevant to Chl-breakdown. These porphyrins are now recognized as abundant "molecular fossils". Most of the known "petro-porphyrins" are Ni(II)- or vanadyl-complexes of a large variety of substituted porphyrins, that carry remnants of the substitution pattern of natural chlorophyll-derivatives; accordingly, some of these are typically associated with a degradation of chlorophyll (see Scheme 19 for a selection of two structural formulae) (*109, 115*). The "petro-porphyrins" are remnants (from partial degradation under anaerobic conditions) of porphyrins or chlorins available and used in the "geological window" of the biosphere and have found use as geochemical biomarkers in petroleum (*109, 115*).

The most visible aspects of Chl-catabolism clearly concern the emergence of the "fall colours" (*4, 7*) and ripening of fruit (*55*), biological phenomena due to higher plants. The factors and conditions responsible for the induction of chlorophyll breakdown in higher plants are still incompletely understood (*116, 117, 118*). Light is an important factor and photo-periodical control operates (*e.g.*) in deciduous trees (*3, 7*). Leaf yellowing and senescence processes including chlorophyll breakdown have demonstrated to be subject to control by phytohormones, and are hastened by ethylene and abscissic acid (*118*). Conversely,

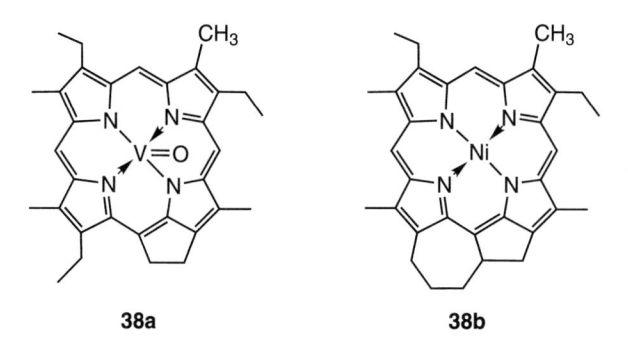

38a   **38b**

Scheme 19. Formulae of two representative "petro-porphyrins". Vanadyl-porphyrinate **38a** and nickel-porphyrinate **38b**

cytokinin inhibits or retards chlorophyll breakdown as well as other senescence processes (*118*). Both phytohormones (cytokinin and abscissic acid) were found to have regulatory effects on PaO (*5, 119*).

Over fifty senescence associated genes in higher plants have been identified (*5, 120*), among them the ones coding for chlorophyllase (*121, 122*), RCC-reductase (*83*), PaO (*79*), and, most recently, Mendel's "green gene" (*29*). "Accelerated cell death genes" (*acd*-1 and *acd*-2) in *Arabidopsis thaliana* were correlated with the absence of functioning RCC-reductase in mutants of this plant (*82*) and with senescence induced Chl-breakdown, as the marker of this visual form of programmed cell death in plants (*6, 81, 123*).

Senescence processes play a very prominent role in the recycling of nutrients, such as reductively fixed nitrogen and magnesium ions from senescent leaves to other parts of the plant (*9*). About one third of the total amount of the reductively fixed nitrogen contained in mature chloroplasts is represented by the proteins of the thylakoid pigment complexes. During senescence, chloroplast proteins are broken down and amino acids are exported for re-use in developing leaves or for the filling of seeds with reserve proteins. However, the apoproteins of chlorophyll are not degraded efficiently as long as the pigments are bound intact, and plant mutants that are disturbed in chlorophyll breakdown (stay-green genotypes) have a metabolic disadvantage due to incomplete nitrogen recycling during senescence (*113, 124*).

At present, there is no evidence of rapid breakdown of Chl beyond the stage of tetrapyrroles. Chl-breakdown, therefore, is not aimed at reusing the four nitrogen atoms of the chlorin macrocycle (which represents only a few percent of total leaf nitrogen) (*1, 2, 3, 4, 5, 6*), but rather at rapidly destroying the chromophores of photoactive Chls. So far, two main consequences of the degradation of Chl were identified: i) the dismantling of Chl protein complexes, as a prerequisite of efficient enzyme catalyzed protein degradation (*113, 124*); ii) the freed Chls are phototoxic and the machinery of Chl catabolism is a vitally important detoxification process. A third consequence may result from a possible physiological role of the NCCs in the plants, as they have recently been found to be effective antioxidants (*55*). Remarkably, NCCs (from degradation of chlorophyll) thus also exhibit similar properties, as antioxidants, as bilirubin (**39**), a reduced form of biliverdin (**19**), the tetrapyrrolic breakdown product of heme, see *e.g.* (*87, 125*), which is important in the metabolism of mammals. Indeed, since the breakdown of protein and the recycling of nutrients depend on a well organized metabolism, it is important for cells to remain viable to the very end of the senescence period.

References, pp. 37–43

The biochemistry of the cleavage of the porphinoid macrocycle by the mono-oxygenase PaO (the "key step" of Chl breakdown in green plants) as well as "fate" and "role" of the tetrapyrrolic breakdown products, the regulation of Chl breakdown steps, are still (largely) unsolved and highly intriguing plant-biological and biochemical questions. Likewise, the chemical reactivity and structural properties of natural Chl catabolites are only revealed to a marginal extent. Research on chlorophyll breakdown is bound to continue providing fascinating and important insights and to allow for further glimpses at the often fascinating interplay of ubiquitous natural products and basic life processes.

Acknowledgements

I am indebted to Philippe Matile for his inspirations to this scientific adventure, and to Stefan Hörtensteiner and his coworkers, for their continued fruitful and stimulating collaboration. I would like to thank specifically Thomas Müller, Simone Moser, Markus Ulrich, and Michael Oberhuber, and the former members of the chlorophyll group, for their contributions to our research. I am grateful to Paula Enders for her excellent technical assistance. Over the years, our work has been supported generously by the Austrian National Science Foundation (FWF), and at present with the project P19596.

References

1. Kräutler B (2003) Chlorophyll Breakdown and Chlorophyll Catabolites. In: Kadish KM, Smith KM, Guilard R (eds.) The Porphyrin Handbook, Vol. 13, p. 183. Elsevier Science, Oxford
2. Kräutler B, Hörtensteiner S (2006) Chlorophyll Catabolites and the Biochemistry of Chlorophyll Breakdown. In: Grimm B, Porra R, Rüdiger W, Scheer H (eds.) Chlorophylls and Bacteriochlorophylls: Biochemistry, Biophysics, Functions and Applications, p. 237. Springer, Dordrecht, The Netherlands
3. Matile P, Hörtensteiner S, Thomas H, Kräutler B (1996) Chlorophyll Breakdown in Senescent Leaves. Plant Physiol **112**: 1403
4. Kräutler B, Matile P (1999) Solving the Riddle of Chlorophyll Breakdown. Acc Chem Res **32**: 35
5. Hörtensteiner S (2006) Chlorophyll Degradation During Senescence. Annu Rev Plant Biol **57**: 55
6. Dangl JL, Dietrich RA, Thomas H (2001) Senescence and Programmed Cell Death. In: Buchanan BB, Gruissem W, Jones RL (eds.) Biochemistry and Molecular Biology of Plants, p. 1044. Am Soc Plant Physiol, Rockville, MD, USA
7. Matile P (2000) Biochemistry of Indian Summer: Physiology of Autumnal Leaf Coloration. Exp Gerontol **35**: 145
8. Brown SB, Houghton JD, Hendry GAF (1991) Chlorophyll Breakdown. In: Scheer H (ed.) Chlorophylls, p. 465. CRC Press, Boca Raton, FL, USA
9. Matile P (1987) Senescence in Plants and Its Importance for Nitrogen-Metabolism. Chimia **41**: 376

10. Kräutler B, Jaun B, Bortlik K, Schellenberg M, Matile P (1991) On the Enigma of Chlorophyll Degradation – The Constitution of a Secoporphinoid Catabolite. Angew Chem Int Ed **30**: 1315
11. Scheer H (ed.) (1991) Chlorophylls. CRC Press, Boca Raton, FL, USA
12. Grimm B, Porra R, Rüdiger W, Scheer H (eds.) (2006) Chlorophylls and Bacteriochlorophylls: Biochemistry, Biophysics, Functions and Applications. Springer, Dordrecht, The Netherlands
13. Montforts FP, Glasenapp-Breiling M (2002) Naturally Occurring Cyclic Tetrapyrroles. In: Herz W, Falk H, Kirby GW (eds.) Progress in the Chemistry of Organic Natural Products, p. 84. Springer, Wien
14. Beale SI, Weinstein JD (1991) Biochemistry and Regulation of Photosynthetic Pigment Formation in Plants and Algae. In: Jordan PM (ed.) Biosynthesis of Tetrapyrroles, p. 155. Elsevier Science, Amsterdam
15. Rüdiger W (2003) The Last Step of Chlorophyll Synthesis. In: Kadish KM, Smith KM, Guilard R (eds.) The Porphyrin Handbook, Vol. 13, p. 71. Elsevier Science, Amsterdam
16. Bollivar DW (2003) Intermediate Steps in Chlorophyll Biosynthesis: Methylation and Cyclization. In: Kadish KM, Smith KM, Guilard R (eds.) The Porphyrin Handbook, Vol. 13, p. 49. Elsevier Science, Amsterdam
17. Ortiz de Montellano PR, Auclair K (2003) Heme Oxygenase Structure and Mechanism. In: Kadish KM, Smith KM, Guilard R (eds.) The Porphyrin Handbook, Vol. 12, p. 183. Academic Press, Amsterdam
18. Woodward RB, Skaric V (1961) A New Aspect of the Chemistry of Chlorins. J Am Chem Soc **83**: 4676
19. Brown SB, Smith KM, Bisset GMF, Troxler RF (1980) Mechanism of Photo-Oxidation of Bacteriochlorophyll-C Derivatives – A Possible Model for Natural Chlorophyll Breakdown. J Biol Chem **255**: 8063
20. Nakamura H, Musicki B, Kishi Y, Shimomura O (1988) Structure of the Light Emitter in Krill (*Euphausia pacifica*) Bioluminescence. J Am Chem Soc **110**: 2683
21. Nakamura H, Kishi Y, Shimomura O, Morse D, Hastings JW (1989) Structure of Dinoflagellate Luciferin and Its Enzymatic and Nonenzymatic Air-Oxidation Products. J Am Chem Soc **111**: 7607
22. Matile P, Ginsburg S, Schellenberg M, Thomas H (1987) Catabolites of Chlorophyll in Senescent Leaves. J Plant Physiol **129**: 219
23. Thomas H, Bortlik K, Rentsch D, Schellenberg M, Matile P (1989) Catabolism of Chlorophyll *in vivo* – Significance of Polar Chlorophyll Catabolites in a Non-Yellowing Senescence Mutant of *Festuca pratensis* Huds. New Phytol **111**: 3
24. Matile P, Ginsburg S, Schellenberg M, Thomas H (1988) Catabolites of Chlorophyll in Senescing Barley Leaves Are Localized in the Vacuoles of Mesophyll-Cells. Proc Natl Acad Sci USA **85**: 9529
25. Bortlik K, Peisker C, Matile P (1990) A Novel Type of Chlorophyll Catabolite in Senescent Barley Leaves. J Plant Physiol **136**: 161
26. Peisker C, Thomas H, Keller F, Matile P (1990) Radiolabeling of Chlorophyll for Studies on Catabolism. J Plant Physiol **136**: 544
27. Kräutler B, Jaun B, Amrein W, Bortlik K, Schellenberg M, Matile P (1992) Breakdown of Chlorophyll – Constitution of a Secoporphinoid Chlorophyll Catabolite Isolated from Senescent Barley Leaves. Plant Physiol Biochem **30**: 333
28. Mendel G (1865) Versuche über Pflanzenhybriden. Verh Naturw Verein Brünn **4**: 3
29. Armstead I, Donnison I, Aubry S, Harper J, Hörtensteiner S, James C, Mani J, Moffet M, Ougham H, Roberts L, Thomas A, Weeden N, Thomas H, King I (2007) Cross-species Identification of Mendel's Locus. Science **315**: 73

30. Willstätter R, Stoll A (1913) Die Wirkungen der Chlorophyllase. Untersuchungen über Chlorophyll, p. 172. Julius Springer, Berlin
31. Spremulli L (2001) Protein synthesis, assembly and degradation. In: Buchanan BB, Gruissem W, Jones RL (eds.) Biochemistry and Molecular Biology of Plants, p. 412. Am. Soc. Plant Physiologists, Rockville, MD, USA
32. Matile P, Schellenberg M, Vicentini F (1997) Localization of Chlorophyllase in the Chloroplast Envelope. Planta **201**: 96
33. Hynninen PH (1991) Chemistry of Chlorophylls: Modifications. In: Scheer H (ed.) Chlorophylls, p. 145. CRC Press, Boca Raton, FL, USA
34. Bachmann A, Fernandez-Lopez J, Ginsburg S, Thomas H, Bouwkamp JC, Solomos T, Matile P (1994) Stay-Green Genotypes of *Phaseolus vulgaris* L. – Chloroplast Proteins and Chlorophyll Catabolites during Foliar Senescence. New Phytol **126**: 593
35. Peisker C, Düggelin T, Rentsch D, Matile P (1989) Phytol and the Breakdown of Chlorophyll in Senescent Leaves. J Plant Physiol **135**: 428
36. Müller T, Moser S, Ongania KH, Pružinska A, Hörtensteiner S, Kräutler B (2006) A Divergent Path of Chlorophyll Breakdown in the Model Plant *Arabidopsis thaliana*. Chem BioChem **7**: 40
37. Ito H, Tanaka Y, Tsuji H, Tanaka A (1993) Conversion of Chlorophyll *b* to Chlorophyll *a* by Isolated Cucumber Etioplasts. Arch Biochem Biophys **306**: 148
38. Ito H, Tanaka A (1996) Determination of the Activity of Chlorophyll *b* to Chlorophyll *a* Conversion During Greening of Etiolated Cucumber Cotyledons by Using Pyrochlorophyllide *b*. Plant Physiol Biochem **34**: 35
39. Scheumann V, Schoch S, Rüdiger W (1999) Chlorophyll *b* Reduction During Senescence of Barley Seedlings. Planta **209**: 364
40. Tanaka A, Ito H, Tanaka R, Tanaka NK, Yoshida K, Okada K (1998) Chlorophyll *a* Oxygenase (CAO) is Involved in Chlorophyll *b* Formation from Chlorophyll *a*. Proc Natl Acad Sci USA **95**: 12719
41. Hörtensteiner S, Vicentini F, Matile P (1995) Chlorophyll Breakdown in Senescent Cotyledons of Rape, *Brassica napus* L. – Enzymatic Cleavage of Phaeophorbide *a* in vitro. New Phytol **129**: 237
42. Folly P, Engel N (1999) Chlorophyll *b* to Chlorophyll *a* Conversion Precedes Chlorophyll Degradation in *Hordeum vulgare* L. J Biol Chem **274**: 21811
43. Shioi Y, Watanabe K, Takamiya K (1996) Enzymatic Conversion of Pheophorbide *a* to the Precursor of Pyropheophorbide *a* in Leaves of *Chenopodium album*. Plant Cell Physiol **37**: 1143
44. Langmeier M, Ginsburg S, Matile P (1993) Chlorophyll Breakdown in Senescent Leaves – Demonstration of Mg-Dechelatase Activity. Physiol Plant **89**: 347
45. Schoch S, Scheer H, Schiff JA, Rüdiger W, Siegelman HW (1981) Pyropheophytin *a* Accompanies Pheophytin *a* in Darkened Light Grown Cells of *Euglena*. Z Naturf C J Biosci **36**: 827
46. Shioi Y, Tatsumi Y, Shimokawa K (1991) Enzymatic Degradation of Chlorophyll in *Chenopodium album*. Plant Cell Physiol **32**: 87
47. Doi M, Inage T, Shioi Y (2001) Chlorophyll Degradation in a *Chlamydomonas reinhardtii* Mutant: An Accumulation of Pyropheophorbide *a* by Anaerobiosis. Plant Cell Physiol **42**: 469
48. Mühlecker W, Kräutler B (1996) Breakdown of Chlorophyll: Constitution of Nonfluorescing Chlorophyll-catabolites from Senescent Cotyledons of the Dicot Rape. Plant Physiol Biochem **34**: 61
49. Iturraspe J, Moyano N, Frydman B (1995) A New 5-Formylbilinone as the Major Chlorophyll *a* Catabolite in Tree Senescent Leaves. J Org Chem **60**: 6664

50. Curty C, Engel N (1996) Chlorophyll Catabolism. 9. Detection, Isolation and Structure Elucidation of a Chlorophyll *a* Catabolite from Autumnal Senescent Leaves of *Cercidiphyllum japonicum*. Phytochem **42**: 1531
51. Oberhuber M, Berghold J, Mühlecker W, Hörtensteiner S, Kräutler B (2001) Chlorophyll Breakdown – On a Nonfluorescent Chlorophyll Catabolite from Spinach. Helv Chim Acta **84**: 2615
52. Berghold J, Breuker K, Oberhuber M, Hörtensteiner S, Kräutler B (2002) Chlorophyll Breakdown in Spinach: On the Structure of Five Nonfluorescent Chlorophyll Catabolites. Photosynth Res **74**: 109
53. Berghold J, Eichmüller C, Hörtensteiner S, Kräutler B (2004) Chlorophyll Breakdown in Tobacco: On the Structure of Two Nonfluorescent Chlorophyll Catabolites. Chem Biodiv **1**: 657
54. Berghold J, Müller T, Ulrich M, Hörtensteiner S, Kräutler B (2006) Chlorophyll Breakdown in Maize: On the Structure of Two Nonfluorescent Chlorophyll Catabolites. Monatsh Chem **137**: 751
55. Müller T, Ulrich M, Ongania KH, Kräutler B (2007) Colorless and Nonfluorescent Chlorophyll Catabolites are Identified in Ripening Fruit and are Effective Antioxidants. Angew Chem Int Ed **46**: 8699
56. Oberhuber M, Berghold J, Breuker K, Hörtensteiner S, Kräutler B (2003) Breakdown of Chlorophyll: A Nonenzymatic Reaction Accounts for the Formation of the Colorless "Nonfluorescent" Chlorophyll Catabolites. Proc Natl Acad Sci USA **100**: 6910
57. Engel N, Jenny TA, Mooser V, Gossauer A (1991) Chlorophyll Catabolism in *Chlorella protothecoides* – Isolation and Structure Elucidation of a Red Bilin Derivative. FEBS Lett **293**: 131
58. Gossauer A, Engel N (1996) Chlorophyll Catabolism – Structures, Mechanisms, Conversions. J Photochem Photobiol B: Biol **32**: 141
59. Engel N, Curty C, Gossauer A (1996) Chlorophyll Catabolism in *Chlorella protothecoides*. 8. Facts and Artefacts. Plant Physiol Biochem **34**: 77
60. Pružinska A, Tanner G, Aubry S, Anders I, Moser S, Müller T, Ongania K-H, Kräutler B, Youn J-Y, Liljegren SJ, Hörtensteiner S (2005) Chlorophyll Breakdown in Senescent *Arabidopsis* Leaves. Characterization of Chlorophyll Catabolites and of Chlorophyll Catabolic Enzymes Involved in the Degreening Reaction. Plant Physiol **139**: 52
61. Scheer H (1991) Structure and Occurrence of Chlorophylls. In: Scheer H (ed.) Chlorophylls, p. 3. CRC Press, Boca Raton, FL, USA
62. Mühlecker W, Ongania KH, Kräutler B, Matile P, Hörtensteiner S (1997) Tracking Down Chlorophyll Breakdown in Plants: Elucidation of the Constitution of a "Fluorescent" Chlorophyll Catabolite. Angew Chem Int Ed **36**: 401
63. Matile P, Schellenberg M, Peisker C (1992) Production and Release of a Chlorophyll Catabolite in Isolated Senescent Chloroplasts. Planta **187**: 230
64. Ginsburg S, Schellenberg M, Matile P (1994) Cleavage of Chlorophyll-Porphyrin – Requirement for Reduced Ferredoxin and Oxygen. Plant Physiol **105**: 545
65. Matile P, Düggelin T, Schellenberg M, Rentsch D, Bortlik K, Peisker C, Thomas H (1989) How and Why Is Chlorophyll Broken down in Senescent Leaves. Plant Physiol Biochem **27**: 595
66. Matile P, Kräutler B (1995) Wie und warum bauen Pflanzen das Chlorophyll *ab*? (How and Why Do Plants Decompose Chlorophyll? Molecular Fundamentals of Leaf Yellowing). Chemie in unserer Zeit **29**: 298

67. Mühlecker W, Kräutler B, Moser D, Matile P, Hörtensteiner S (2000) Breakdown of Chlorophyll: A Fluorescent Chlorophyll Catabolite from Sweet Pepper (*Capsicum annuum*). Helv Chim Acta **83**: 278
68. Rodoni S, Vicentini F, Schellenberg M, Matile P, Hörtensteiner S (1997) Partial Purification and Characterization of Red Chlorophyll Catabolite Reductase, a Stroma Protein Involved in Chlorophyll Breakdown. Plant Physiol **115**: 677
69. Rodoni S, Mühlecker W, Anderl M, Kräutler B, Moser D, Thomas H, Matile P, Hörtensteiner S (1997) Chlorophyll Breakdown in Senescent Chloroplasts. Cleavage of Pheophorbide *a* in Two Enzymic Steps. Plant Physiol **115**: 669
70. Vicentini F, Hörtensteiner S, Schellenberg M, Thomas H, Matile P (1995) Chlorophyll Breakdown in Senescent Leaves – Identification of the Biochemical Lesion in a Stay-Green Genotype of *Festuca pratensis* Huds. New Phytol **129**: 247
71. Schellenberg M, Matile P, Thomas H (1993) Production of a Presumptive Chlorophyll Catabolite *in-vitro* – Requirement for Reduced Ferredoxin. Planta **191**: 417
72. Hörtensteiner S, Wüthrich KL, Matile P, Ongania KH, Kräutler B (1998) The Key Step in Chlorophyll Breakdown in Higher Plants – Cleavage of Pheophorbide *a* Macrocycle by a Monooxygenase. J Biol Chem **273**: 15335
73. Curty C, Engel N, Gossauer A (1995) Evidence for a Monooxygenase-Catalyzed Primary Process in the Catabolism of Chlorophyll. FEBS Lett **364**: 41
74. Iturraspe J, Gossauer A (1992) A Biomimetic Partial Synthesis of the Red Chlorophyll *a* Catabolite from *Chlorella prototheicoides*. Tetrahedron **48**: 6807
75. Gossauer A (2003) Synthesis of Bilins. In: Kadish KM, Smith KM, Guilard R (eds.) The Porphyrin Handbook, Vol. 13, p. 237. Academic Press/Elsevier, Amsterdam
76. Kräutler B, Mühlecker W, Anderl M, Gerlach B (1997) Breakdown of Chlorophyll: Partial Synthesis of a Putative Intermediary Catabolite – Preliminary Communication. Helv Chim Acta **80**: 1355
77. Topalov G, Kishi Y (2001) Chlorophyll Catabolism Leading to the Skeleton of Dinoflagellate and Krill Luciferins: Hypothesis and Model Studies. Angew Chem Int Ed **40**: 4010
78. Gossauer A (1994) Catabolism of Tetrapyrroles. Chimia **48**: 352
79. Pružinska A, Tanner G, Anders I, Roca M, Hörtensteiner S (2003) Chlorophyll Breakdown: Pheophorbide a Oxygenase is a Rieske-type Iron-sulfur Protein, Encoded by the *Accelerated Cell Death 1* Gene. Proc Natl Acad Sci USA **100**: 15259
80. Pružinska A, Anders I, Aubry S, Schenk N, Tapernoux-Lüthi E, Müller T, Kräutler B, Hörtensteiner S (2007) *In Vivo* Participation of Red Chlorophyll Catabolite Reductase in Chlorophyll Breakdown and in Detoxification of Photodynamic Chlorophyll Catabolites. Plant Cell **19**: 369
81. Greenberg JT, Guo AL, Klessig DF, Ausubel FM (1994) Programmed Cell-Death in Plants – A Pathogen-Triggered Response Activated Coordinately with Multiple Defense Functions. Cell **77**: 551
82. Mach JM, Castillo AR, Hoogstraten R, Greenberg JT (2001) The *Arabidopsis* Accelerated Cell Death Gene ACD2 Encodes Red Chlorophyll Catabolite Reductase and Suppresses the Spread of Disease Symptoms. Proc Natl Acad Sci USA **98**: 771
83. Wüthrich KL, Bovet L, Hunziker PE, Donnison IS, Hörtensteiner S (2000) Molecular Cloning, Functional Expression and Characterisation of RCC Reductase Involved in Chlorophyll Catabolism. Plant J **21**: 189
84. Hörtensteiner S, Rodoni S, Schellenberg M, Vicentini F, Nandi OI, Qui YL, Matile P (2000) Evolution of Chlorophyll Degradation: The Significance of RCC Reductase. Plant Biol **2**: 63

85. Matile P, Hörtensteiner S, Thomas H (1999) Chlorophyll Degradation. Ann Rev Plant Physiol Plant Mol Biol **50**: 67
86. Oberhuber M, Kräutler B (2002) Breakdown of Chlorophyll: Electrochemical Bilin Reduction Provides Synthetic Access to Fluorescent Chlorophyll Catabolites. Chem BioChem **3**: 104
87. Falk H (1989) Chemistry of Linear Oligopyrroles and Bile Pigments. Springer, Wien New York
88. Cornejo J, Beale SI (1997) Phycobilin Biosynthetic Reactions in Extracts of Cyanobacteria. Photosynth Res **51**: 223
89. Frankenberg N, Lagarias JC (2003) Biosynthesis and Biological Functions of Bilins. In: Kadish KM, Smith KM, Guilard R (eds.) The Porphyrin Handbook, Vol. 13, p. 211. Elsevier Science, Oxford, UK
90. Frankenberg N, Mukougawa K, Kohchi T, Lagarias JC (2001) Functional Genomic Analysis of the HY2 Family of Ferredoxin-Dependent Bilin Reductases from Oxygenic Photosynthetic Organisms. Plant Cell Physiol **13**: 965
91. Eschenmoser A (1988) Vitamin-B_{12} – Experiments Concerning the Origin of Its Molecular-Structure. Angew Chem Int Ed **27**: 5
92. Oberhuber M, Berghold J, Kräutler B (2008) Chlorophyll Breakdown by Bio-mimetic Synthesis. Angew Chem Int Ed **47**: 3057
93. Ginsburg S, Matile P (1993) Identification of Catabolites of Chlorophyll-Porphyrin in Senescent Rape Cotyledons. Plant Physiol **102**: 521
94. Hörtensteiner S, Kräutler B (2000) Chlorophyll Breakdown in Oilseed Rape. Photosynth Res **64**: 137
95. Mühlecker W, Kräutler B, Ginsburg S, Matile P (1993) Breakdown of Chlorophyll – A Tetrapyrrolic Chlorophyll Catabolite from Senescent Rape Leaves. Helv Chim Acta **76**: 2976
96. Müller T, Kräutler B (unpublished)
97. Hörtensteiner S (1998) NCC Malonyltransferase Catalyses the Final Step of Chlorophyll Breakdown in Rape (*Brassica napus*). Phytochem **49**: 953
98. Matile P (1997) The Vacuole and Cell Senescence. In: Callow JA (ed.) Advances in Botanical Research, p. 87. Academic Press, New York
99. Harborne JB (1986) The Natural Distribution in Angiosperms of Anthocyanins Acylated with Aliphatic Dicarboxylic-Acids. Phytochem **25**: 1887
100. Hinder B, Schellenberg M, Rodoni S, Ginsburg S, Vogt E, Martinoia E, Matile P, Hörtensteiner S (1996) How Plants Dispose of Chlorophyll Catabolites – Directly Energized Uptake of Tetrapyrrolic Breakdown Products into Isolated Vacuoles. J Biol Chem **271**: 27233
101. Losey FG, Engel N (2001) Isolation and Characterization of a Urobilinogenoidic Chlorophyll Catabolite from *Hordeum vulgare*. J Biol Chem **276**: 8643
102. Llewellyn CA, Mantoura RFC, Brereton RG (1990) Products of Chlorophyll Photodegradation. 2. Structural Identification. Photochem Photobiol **52**: 1043
103. Suzuki Y, Shioi Y (1999) Detection of Chlorophyll Breakdown Products in the Senescent Leaves of Higher Plants. Plant Cell Physiol **40**: 909
104. Oshio Y, Hase E (1969) (1) Studies on Red Pigments Excreted by Cells of *Chlorella protothecoides* During Process of Bleaching Induced by Glucose or Acetate. I. Chemical Properties of Red Pigments. Plant Cell Physiol **10**: 41
105. Oshio Y, Hase E (1969) Studies on Red Pigments Excreted by Cells of *Chlorella protothecoides* During Process of Bleaching Induced by Glucose or Acetate. 2. Mode of Formation of Red Pigments. Plant Cell Physiol **10**: 51

106. Iturraspe J, Engel N, Gossauer A (1994) Chlorophyll Catabolism. 5. Isolation and Structure Elucidation of Chlorophyll *b* Catabolites in *Chlorella protothecoides*. Phytochem **35**: 1387
107. Curty C, Engel N (1997) Chlorophyll Catabolism: High Stereoselectivity in the Last Step of the Primary Ring Cleaving Process. Plant Physiol Biochem **35**: 707
108. Raven HP, Evert RF, Eichhorn SE (1987) Biology of Plants. Worth Publishers, New York
109. Morel A (2006) Meeting the Challenge of Monitoring Chlorophyll in the Ocean from Outer Space. In: Grimm B, Porra R, Rüdiger W, Scheer H (eds.) Chlorophylls and Bacteriochlorophylls Biochemistry, Biophysics, Functions and Applications, p. 521. Springer, Dordrecht
110. Dunlap JC, Hastings JW, Shimomura O (1981) Dinoflagellate Luciferin is Structurally Related to Chlorophyll. FEBS Lett **135**: 273
111. Stojanovic MN, Kishi Y (1994) Dinoflagellate Bioluminescence – The Chromophore of Dinoflagellate Luciferin. Tetrahedron Lett **35**: 9343
112. Stojanovic MN, Kishi Y (1994) Dinoflagellate Bioluminescence – Chemical Behavior of the Chromophore Towards Oxidants. Tetrahedron Lett **35**: 9347
113. Thomas H, Ougham H, Hörtensteiner S (2001) Recent Advances in the Cell Biology of Chlorophyll Catabolism. Adv Bot Res **35**: 1
114. Treibs A (1936) Chlorophyll and Heme Derivatives in Organic Mineral Materials. Angew Chem **49**: 682
115. Callot HJ, Ocampo R (2000) Geochemistry of Porphyrins. In: Kadish KM, Smith KM, Guilard R (eds.) The Porphyrin Handbook, Vol. 1, p. 349. Elsevier Science, Oxford
116. Matile P (2001) Senescence and Cell Death in Plant Development: Chloroplast Senescence and Its Regulation. In: Aro E-M, Andersson B (eds.) Regulation of Photosynthesis, p. 277. Kluwer Academic Publishers, Dordrecht
117. Hörtensteiner S (1999) Chlorophyll Breakdown in Higher Plants and Algae. Cell Mol Life Sci **56**: 330
118. Noodén LA, Leopold AC (eds.) (1988) Senescing and Aging in Plants. Academic Press, San Diego
119. Rodoni S, Schellenberg M, Matile P (1998) Chlorophyll Breakdown in Senescing Barley Leaves as Correlated with Phaeophorbide *a* Oxygenase Activity. J Plant Physiol **152**: 139
120. Smart CM (1994) Gene-Expression During Leaf Senescence. New Phytol **126**: 419
121. Jacob-Wilk D, Holland D, Goldschmidt EE, Riov J, Eyam Y (1999) Chlorophyll Breakdown by Chlorophyllase: Isolation and Functional Expression of the Chlase1 Gene from Ethylene-treated Citrus Fruit and Its Regulation During Development. Plant J **20**: 653
122. Tsuchiya T, Ohta H, Okawa K, Iwamatsu A, Shimada H, Masuda T, Takamiya K-I (1999) Cloning of Chlorophyllase, the Key Enzyme in Chlorophyll Degradation: Finding of a Lipase Motif and the Induction by Methyl Jasmonate. Proc Natl Acad Sci USA **96**: 15362
123. Thomas H, Ougham HJ, Wagstaff C, Stead AD (2003) Defining Senescence and Death. J Exp Bot **54**: 1127
124. Thomas H (1997) Chlorophyll: A Symptom and a Regulator of Plastid Development. New Phytol **136**: 163
125. Stocker R, Yamamoto Y, McDonagh AF, Glazer AN, Ames BN (1987) Bilirubin is an Antioxidant of Possible Physiological Importance. Science **235**: 1043

Steroidal Saponins

N. P. Sahu*, *S. Banerjee, N. B. Mondal*, and *D. Mandal*

Indian Institute of Chemical Biology, 4 Raja S C Mullick Road, Kolkata 700 032, India

Contents

1. Introduction	45
2. Isolation	46
3. Structure Elucidation	49
3.1. Conventional Methods	50
3.2. Spectrometry Coupled with Chemical Methods	52
3.3. Modern Spectrometric Methods	55
3.3.1. Mass Spectrometry	55
3.3.2. NMR Spectroscopy	57
3.3.2.1. ^{1}H NMR Spectroscopy	57
3.3.2.2. ^{13}C NMR Spectroscopy	58
3.3.2.3. 2D NMR Spectroscopy	59
4. Biological Activity	62
4.1. Cytotoxic Activity Against Cancer Cell Lines	63
4.2. Antifungal Activity	66
4.3. Miscellaneous Effects	68
5. Biosynthesis of Steroidal Glycosides	69
6. Report of New Steroidal Saponins (1998–Mid-2006)	70
7. Conclusion	126
Acknowledgement	126
References	127

1. Introduction

The medicinal activities of plants are generally due to the secondary metabolites (*1*) which often occur as glycosides of steroids, terpenoids, phenols, *etc*. Saponins are a group of naturally occurring plant glycosides, characterized by their strong foam-forming properties in aqueous solution. The cardiac glycosides also possess this property but are

---

* E-mail: npsahu@iicb.res.in, Fax: +91-33-2473 5197

classified separately because of their specific biological activity. Unlike the cardiac glycosides, saponins generally do not affect the heart. These are classified as steroid or triterpenoid saponins depending on the nature of the aglycone. Steroidal glycosides are naturally occurring sugar conjugates of C_{27} steroidal compounds. The aglycone of a steroid saponin is usually a spirostanol or a furostanol. The glycone parts of these compounds are mostly oligosaccharides, arranged either in a linear or branched fashion, attached to hydroxyl groups through an acetal linkage (2, 3). Another class of saponins, the basic steroid saponins, contain nitrogen analogues of steroid sapogenins as aglycones.

Steroidal glycosides have drawn much attention in the last few decades not only as economically important raw materials for the pharmaceutical industry used in the production of various steroidal hormones (4–7) but also as biologically active compounds (8–13) and as ingredients for cosmetics (14). General reviews dealing with steroid saponins have been published earlier by Tschesche and Wulff (15), Elks (16, 17) and Takeda (18). Following our previous review of steroid saponins (19) which covered the literature up to 1980 a number of reviews dealing with specific aspects of spirostanes, furostanes and their glycosides have appeared (20–32) covering the literature up to early 1998. The present review is a compilation of steroid saponins isolated during the period 1998 to mid 2006 together with their biological activities. It also includes a summary of the latest developments in purification processes and structure elucidation techniques (mainly NMR and mass spectrometry).

2. Isolation

The methods for isolation of steroid saponins are similar to those of triterpenoid saponins. Since glycosides, as a class, are particularly prone to enzymatic or microbial degradation, processing of plant material needs to be started soon after collection to avoid delays. Air-dried powdered plant material is defatted and then extracted, either with cold or hot methanol or ethanol or with 50% aqueous ethanol or methanol at ambient temperature. Usually the extract is concentrated at reduced pressure, macerated with water, and partitioned successively using ethyl acetate and n-BuOH. Most of the saponin constituents are found in the n-BuOH soluble fraction. However, highly polar glycosides may be found in the aqueous layer.

Steroidal saponins are usually highly polar compounds occurring as complex mixtures, and their separation into individual components is a formidable task. The traditional purification and separation process for steroidal saponins consists of repetitive chromatography on silica gel columns using chloroform-methanol and/or chloroform-methanol-water as eluent, followed by fractional crystallization, preparative TLC, *etc*. This method is widely used even today, to get rid of coloring matters and other non-saponin constituents. In this way our group was able to separate five steroidal saponins, kallstroemins A–E, from the aerial parts of *Kallstroemia pubescens* (*33*) and six steroidal saponins, floribundasaponins A–F, from the yams of *Dioscorea floribunda* (*34*). Although such a process may yield homogeneous compounds in a few cases, this classical method is now used mainly for separating the crude saponin mixture into different fractions according to their polarity, final purification being achieved by modern chromatographic techniques. Nowadays the crude saponin mixture is applied to a column of Diaion HP-20, which is washed with water-methanol in various ratios (0, 30, 50, 80 and 100% methanol). Often the saponin fraction is obtained from the 70–100% methanol eluates. Fractions found to have the same pattern on TLC are combined and further purified by silica gel column chromatography (chloroform-methanol/chloroform-methanol-water in various ratios), ODS medium pressure LC and finally by HPLC. Isolation of racemosides A–C, steroidal saponins from the fruits of *Asparagus racemosus* (*35*), may be taken as an example. The air-dried powdered fruits of *Asparagus racemosus* were first defatted at room temperature with petroleum ether and extracted with methanol at ambient temperature. The methanol extract was concentrated under reduced pressure and partitioned between *n*-butanol and water. The organic layer was washed with water and concentrated to dryness under reduced pressure. The residue was applied to a column of Diaion HP-20 and washed with water followed by 30%, 50%, 80% and 100% of methanol. Fractions eluted with 50% methanol contained saponin(s). Repeated chromatographic purification over a silica gel column furnished racemoside A and a mixture of racemosides B and C, which were successfully separated into individual saponins by preparative TLC over silica gel (*35*).

Another example describes the isolation of three hexasaccharides (**I**, **III** and **V**) and two trisaccharides (**II** and **IV**) from the leaves of *Agave fourcroydes* (*36*). Air-dried leaves were extracted with methanol and applied to a column of Diaion HP-20. The fraction eluted with 100% methanol was partitioned between ethyl acetate and 10% aqueous methanol and the aqueous phase was further extracted with *n*-butanol. This *n*-butanol-soluble fraction was subjected to ODS column chromatography and eluted with gradient mixtures of 50–100% methanol in water. Rechromatography over ODS of the 80% methanol-eluted fraction using gradient mixtures of 60–75% methanol in water furnished a fraction eluted with 70–75% methanol. This was further purified by HPLC over an ODS column (Develosil ODS HG-5, 10 × 250 mm, eluent: 75% methanol) to afford **I** and **IV**. The fractions of the *n*-butanol soluble part, eluted in the earlier chromatography with 70 and 80% methanol furnished **II**, **III** and **V** on further purification by ODS column chromatography, Sephadex LH-20 and repeated HPLC.

I. $R_1 = OH$, $R_2 = H_2$
III. $R_1 = H$, $R_2 = O$
V. $R_1 = H$, $R_2 = H_2$

II. $R = OH$
IV. $R = H$

In another example six steroidal glycosides possessing antineoplastic activity were isolated from the African plant *Sansevieria ehrenbergii* following bioactivity-directed isolation procedure. The dried and chipped plant was extracted with methanol-methylene chloride (1:1) at ambient temperature. The extract was separated into methylene chloride and methanol-water phase by addition (30 vol%) of water. The methanol-water extract was fractionated into *n*-hexane, methylene chloride, ethyl acetate, *n*-butanol and aqueous fractions. The respective extracts were repeatedly chromatographed over Sephadex LH-20 and silica gel columns. Finally the streroidal glycosides, sansevierin A, sansevistatins 1 and 2 as well as three known steroidal saponins were separated by HPLC using a Zorbax SB C_{18} column (25 cm × 4.6 mm, 5 µm) with an isocratic mobile phase: 75% methanol in water (*37*).

Sautour *et al.* (*38*) isolated three anti-fungal steroidal saponins from the roots of *Smilax medica*. Dried, powdered roots were boiled thrice with methanol:water (7:3). After filtration the combined extract was evaporated to dryness. The residue was suspended in water and partitioned successively with *n*-hexane and *n*-butanol. The *n*-butanol fraction was evaporated to dryness, dissolved in methanol, concentrated, and the glycosides were precipitated with repeated addition of diethyl ether. The precipitate was filtered, dried and further purified by vacuum-liquid chromatography (VLC) on a C_{18} reversed-phase column using water, various mixtures of water-methanol and finally pure methanol to give four different fractions. The fractions were submitted to MPLC over silica gel (15–40 µm) and eluted with chloroform:methanol:water (13:7:2, lower phase) to give four homogeneous steroidal saponins.

22,25-Epoxy-furost-5-ene and 20,22-*seco*-type steroidal glycosides were isolated from the fruits of *Solanum abutiloides* (*39*) as follows: The fresh fruits were extracted with methanol at room temperature for three months. The dried methanol extract was partitioned between equal volumes of chloroform and water. The aqueous part was dried and subjected to column chromatography over MCI gel CHP20P with methanol-water in various ratios (40, 50, 60, 70, 80 and 90%) to afford different fractions. The fractions were further purified by repeated ODS and silica gel column chromatography using various solvent systems, followed by HPLC on a ODS (PrePAK-500/C_{18}, Waters) column to afford abutilosides L, M, N and O, aculeatisides A and B, and a pregnane-type glycoside, compound Pd.

In another example Zhang *et al.* (*40*) isolated ten furostanol saponins as five pairs of 25*R*- and 25*S*-epimers from the fresh rhizomes of *Polygonatum kingianum*. The fresh rhizomes were extracted thrice with 50% aqueous ethanol. The combined extract was concentrated under reduced pressure, passed over a macroporous resin AB-8 and eluted with a gradient mixture of acetone-water (1:9, 1:1, 4:1) to give three fractions. The fraction eluted with 50% acetone-water was rechromatographed on macroporous resin SP825 and eluted with a gradient mixture of acetone:water (1:4, 3:7, 2:3, 4:1). Further purification of the fraction eluted with acetone:water (3:7) over silica gel (50 µm) and repeated preparative HPLC furnished all the ten homogeneous epimers.

3. Structure Elucidation

Structure determinations of the homogeneous saponins are usually carried out by a combination of chemical and spectroscopic methods. Extensive investigations of the aglycones demonstrated that most of them are spirostane derivatives or modified spirostanes. Furostanol glycosides have also been isolated and characterized, which according to Marker and Lopez (*41*) are precursors of sugar conjugates of spirostanes.

Steroidal glycosides possessing a furostane skeleton do not exhibit IR absorptions at 918 and 900 cm^{-1} characteristic of spirostane derivatives (*42*). Moreover, furostanol glycosides, with some exceptions (*43*), show a characteristic red color on thin layer chromatographic (TLC) plates when developed with *p*-dimethylaminobenzaldehyde in methanol and exposed to hydrochloric acid [Ehrlich reagent (*44*)]. Confirmation of the furostanol structure may also be obtained by analyzing the products obtained by either Marker's degradation or Baeyer-Villiger oxidation followed by hydrolysis (*34, 45*).

18-Norspirostanol derivatives, which possess unusual steroid skeletons with α,β-unsaturated ketone and hydroxyl groups at C-23 and C-24, have been isolated from three Liliaceae plants, *Trillium kamtschaticum, T. tschonoskii* and *Paris quadrifolia* (*46–53*). In a rare case, Yokosuka *et al.* (*54*) have isolated two new steroidal glycosides possessing 3,5-cyclospirostanol and furostanol as the aglycones. However, these new glycosides usually vary only in the carbohydrate chain and the nature of the sugar sequence.

Generally the sugar moieties of steroidal saponins are oligosaccharides consisting of two to five kinds of sugar units. D-Glucose, D-galactose, D-xylose, L-arabinose and L-fucose occur widely, while D-apiose and D-quinovose occur only rarely. Steroidal saponins linked to a 2-deoxyribose unit have also been reported (*55*). The carbohydrate moiety is linked to the aglycone through hydroxyl groups either in a linear or branched fashion.

Structural studies of the saponins can be broadly divided into three stages, *viz*. conventional methods, spectrometry coupled with chemical methods and modern spectrometric methods. With the advent of modern spectroscopic methods, examination of the intact glycoside itself may lead to determination of the complete structure.

3.1. Conventional Methods

The conventional method of structure elucidation of steroidal saponins starts with acid hydrolysis of the homogeneous saponin leading to identification of the aglycone and the individual monosaccharide constituents separately. The structure of the sugar moieties of the glycosides is then determined by identification of the monosaccharides (obtained on acid hydrolysis) by PC, GLC (alditol acetates/TMS derivatives), and HPLC (comparison with authentic samples). Sometimes microhydrolysis is used to identify the monosaccharide constituents (*56*). The method has been applied to the identification of monosaccharide constituents of saponins isolated from *Polycarpon succulentum* (*57, 58*). The saponins were applied to silica gel TLC plates and left in a HCl atmosphere in an oven at 100°C for one hour. After elimination of HCl vapour, authentic sugars were applied to the chromatography plate and developed. The spots were visualized by spraying with anisaldehyde and sulfuric acid followed by heating. The monosaccharides were identified on comparing the spots with those of authentic samples. However, partial hydrolysis or controlled hydrolysis followed by isolation and characterization of prosapogenins and, where possible, by characterization of oligosaccharides is sometimes employed for the determination of the sugar sequence (*59–61*). Mimaki *et al.* (*62*) carried out partial hydrolysis with 0.2 M HCl (dioxane:water, 1:1) at 100°C for two hours to obtain apiose, present as the terminal sugar moiety of steroidal glycosides isolated from *Chlorophytum comosum*.

In some steroidal glycosides, an acyl function is present as part of the sugar moieties. Treatment with sodium methoxide or ammonia solution in methanol at room temperature was found to be suitable for deacylation. Mimaki *et al.* (*63*) have used 10% ammonia solution in methanol to cleave the acetyl group present at C-4 of galactose, keeping the C-6 acetyl function intact, in the structure elucidation of steroidal glycosides isolated from *Ruscus aculeatus*. However, use of 3% sodium methoxide in methanol cleaved both the acetyl groups. Very rarely, a sulfate group is present in the oligosaccharide part of the glycosides. Desulfonication is usually done by solvolysis (*64*). A spirostanol saponin isolated from the underground parts of *Ruscus aculeatus* was desulfonicated by refluxing with a mixture of pyridine and dioxane (*65*). After completion of the reaction, the mixture was passed through a Sep-Pak C_{18} cartridge (Waters) and eluted successively with water and methanol. The fraction eluted with methanol was chromatographed on silica gel to yield the desulfonicated compound. When the aqueous phase was examined by paper chromatography, sulfuric acid was detected as a light yellow spot after spraying the paper with a solution of barium chloride followed by potassium rhodizonate.

β-Glucosidases are usually employed to hydrolyze the β-glucosidic linkage(s) of a glucoside. There are a number of β-glucosidases possessing specific activity for various substrates, such as cyanogenic glucosides (*66, 67*), hydroxamic acid glucosides (*68*), β-linked oligoglucosides (*69, 70*), isoflavonoid glucosides (*71*) or furostanol glycosides (*72–74*). The precursors of spirostane glucosides are furostanol glucosides, in which a glucose unit is linked to the C-26 hydroxyl of the sapogenin. Usually the glucose unit of the latter is cleaved by a β-glucosidase enzyme to form the spiroketal ring of steroidal glycosides.

Thus, the furostanol glycoside 26-O-β-glucosidase from the rhizomes of *Costus speciosus* cleaves the furostanol glycosides protodioscin and protogracillin to the corresponding spirostanes (74).

β-Glucosidase has also been used by other authors (75–79) to cleave the C-26 glucose unit of 26-O-furostanol glucosides. Other enzymes used to cleave the monosaccharide unit of furostanol glycosides are almond emulsin (80) and hesperidinase (81).

The points of attachment of different sugar units are revealed by permethylation of the glycoside followed by hydrolysis or methanolysis and identification of the partially methylated sugar derivatives by GLC. The non-methylated sites of the hydrolysis products of the permethylate revealed the sites of the linkages. Moreover, Smith degradation as well as periodate oxidation have also been employed for determining the nature of the sugar chain in saponins (82–84). One of the most important procedures for determining interglycosidic linkages is to carry out a GC-MS analysis of the derivatised sugars of the permethylated saponins. The permethylated saponin is hydrolyzed, reduced and subsequently acetylated, thus producing the corresponding monosaccharide derivatives, which are analyzed and compared with the data from authentic specimens (85).

The absolute configuration of the monosaccharides can be determined by analyzing the sugars (obtained from hydrolysis experiments) on a chiral HPLC column (86, 87). The absolute stereochemistry of the monosaccharides may also be derived by chiral GC analysis (88, 89). Moreover, one can also compare the observed value of the molecular rotation with the value calculated on the basis of Klyne's rule (90).

3.2. Spectrometry Coupled with Chemical Methods

Since 1980, the advent of modern spectrometric methods like soft ionization mass spectrometry and FT-NMR has made the structural study of saponins somewhat easier. FDMS and FABMS turned out to be very powerful tools in the structure elucidation of saponins. They not only provide the correct molecular weight but also in many instances the sequence of the glycone part. FABMS in conjunction with ^{1}H and ^{13}C NMR spectroscopy (glycosidation and esterification shift rules, comparison of NMR data, utilization of J_{H1H2} values for determining anomeric configurations) together with chemical strategies has simplified structure elucidation of even complicated saponins. Such techniques were successfully applied by us to a number of saponins to determine the molecular weight and the monosaccharide sequence, as well as to assign the carbon and proton resonances by comparison with data of similar struc-

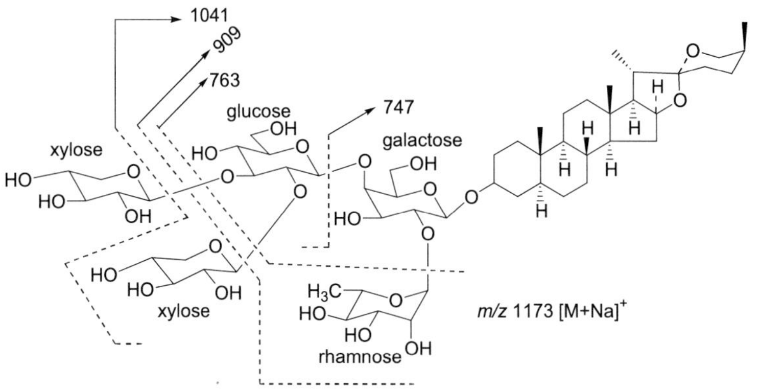

Fig. 1. Characteristic fragments obtained from [M + Na]⁺ ion in the FDMS of tribulosin (**VI**)

tures (*91–94*), using chemical-shift (*95*) and glycosidation shift rules (*93*, *96–98*). The following exemplifies application of both chemical and spectral methods in determining the structure of the steroidal saponin tribulosin (**VI**) isolated from *Tribulus terrestris* (*91*) where FDMS, ^{1}H and ^{13}C NMR, and chemical transformations of tribulosin were successfully utilized. FDMS exhibited ion peaks at *m/z* 1189 and 1173 corresponding to [M + K]⁺ and [M + Na]⁺ respectively thus indicating a molecular weight of **VI** as 1150. Appearance of doubly charged ion peaks corresponding to the ions [M + 2Na]⁺⁺ and [(M + 2Na + H) − xylose]⁺⁺ also supported the molecular weight assignment. The formation of different ion peaks in the FDMS spectrum (Fig. 1) of **VI** indicated that xylose was present as a terminal sugar. Controlled acid hydrolysis (0.75 M H$_2$SO$_4$ in EtOH on a steam bath for 20 min) of tribulosin (**VI**) furnished two prosapogenins (**VII**) and (**VIII**) (Fig. 2). Further acid hydrolysis of the less polar one **VII** yielded neotigogenin (**IX**) as the aglycone and D-galactose as the monosaccharide constituent, indicating that D-galactose was linked directly to the aglycone. The other prosapogenin (**VIII**) on acid hydrolysis furnished two monosaccharides, glucose and galactose. Moreover, treatment of **VI** with sodium metaperiodate followed by acid hydrolysis afforded glucose and galactose as the monosaccharides, indicating that the monosaccharides were interlinked in such a fashion that none of the two sugars had vicinal hydroxyl groups. From the foregoing evidence it was presumed that tribulosin (**VI**) had one of the structures (i), (ii), or (iii). Permethylation and methanolysis liberated methyl 2,3,4-tri-*O*-methyl-L-rhamnopyranoside, methyl 2,3,4-tri-*O*-methyl-D-xylopyranoside, methyl 3,6-di-*O*-methyl-D-galactopyranoside and methyl 4,6-di-*O*-methyl-D-glucopyranoside,

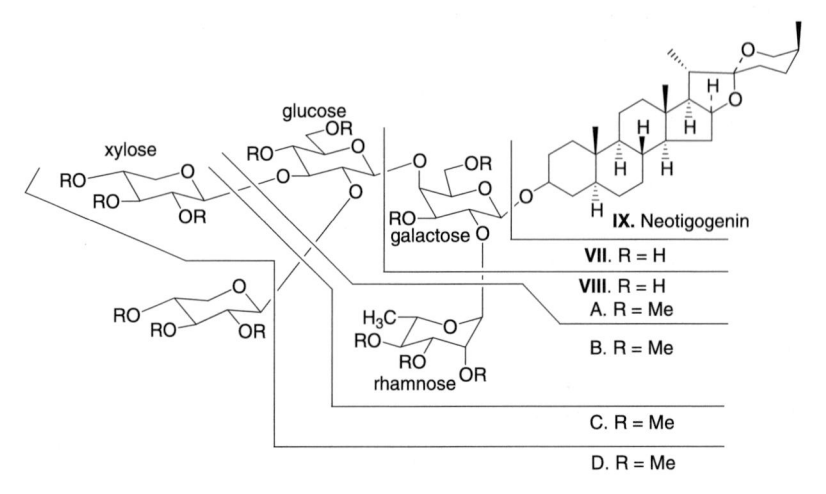

Fig. 2. Controlled hydrolysis and permethylated products of **VI**

identified from GLC studies. Thus tribulosin could be represented by

(i) Neotigogenin — galactose — glucose — xylose
              |
              rhamnose — xylose

(ii) Neotigogenin — galactose — glucose ⟨ xylose / xylose
              |
              rhamnose

(iii) Neotigogenin — galactose — glucose — xylose — xylose
              |
              rhamnose

either of the two isomeric structures (iia) and (iib). The complete structure of tribulosin (**VI**) was established as (iia) through partial hydrolysis followed by permethylation, separation of the partially methylated prosapogenins **A–D** (Fig. 2) and identification of the methanolysis products of tribulosin by GLC (*91*).

(iia) Neotigogenin — galactose^4 — glucose^2 ⟨ xylose / xylose
                   ^{2}| ^{3}
                   rhamnose

(iib) Neotigogenin — galactose^2 — glucose^2 ⟨ xylose / xylose
                   ^{4}| ^{3}
                   rhamnose

References, pp. 127–141

3.3. Modern Spectrometric Methods

With the advent of modern spectroscopic methods, especially 2D-NMR and soft ionization mass spectrometry, the structural study of saponins no longer necessitated most of the time consuming and sample demanding chemical methods. Determination of the structure of saponins by spectrometric methods has the advantage of confining the analysis to the saponin itself, avoiding processing that might produce artefacts. In many cases a complete structure determination is possible by NMR spectroscopy using only a few milligrams of sample (*99*). Furthermore, the recent introduction of HPLC coupled either to a UV photodiode array detector (LC-DAD-UV) and a mass spectrometer, or to an NMR spectrometer (LC-NMR) provides on-line useful structural information of plant constituents with only a minute amount of plant material (*100*).

3.3.1. Mass Spectrometry

Mass spectrometry can provide information not only about the molecular weight and the molecular formula, but also about the number of monosaccharides, and sometimes even the sequence of the oligosaccharide chain. The use of soft-ionization mass spectrometric methods, viz. FAB-MS (*101–106*), field desorption (*107, 108*), plasma desorption (*109, 110*), and laser desorption (*111*) has been extensively discussed by others. In recent years MALDI-TOF-MS and ESI-MS have become popular for structural studies of complex molecules (*112–116*). The use of MALDI-TOF-MS for structural studies of saponins is so far limited to a report (*117*) on the BSA conjugate of the saponin aculeatiside A, where the technique was applied only to determine the ratio of hapten in the antigen conjugate. However, electrospray ionization (ESI) in conjunction with multi-stage tandem mass spectrometry has been shown to constitute a powerful tool for the analysis of saponin mixtures, which can obviate the isolation of individual saponins and provide considerable structural information on various types of compounds including steroidal glycosides isolated from natural sources (*118–124*). Useful information can be obtained by separating the individual parent ions followed by collision-induced decomposition (CID) and analysis of the different fragments. Li *et al.* (*125*) have applied this technique to elucidate the structures of 13 steroid saponins extracted from the rhizomes of *Dioscorea panthaica*. In order to study the fragmentation pathways of these steroid saponins, they also carried out ESI-QTOF-MS/MS of ten authentic steroid saponins. In addition, they have used atmospheric pressure chemical ionization mass spectrometry combined with ion trap

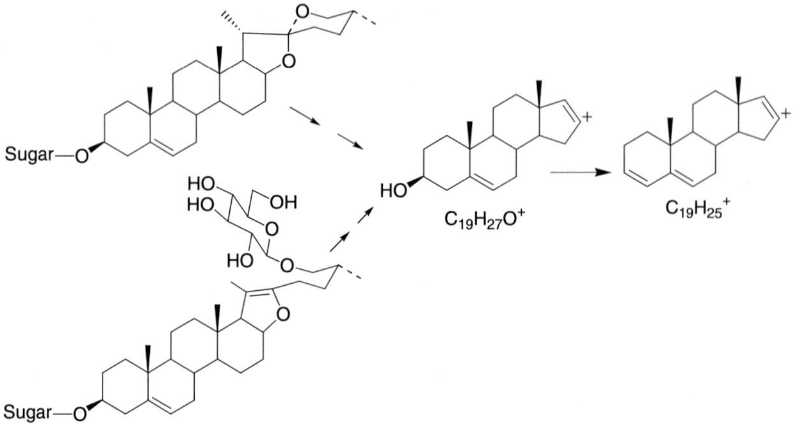

Fig. 3. Diagnostic fragment ions for the spirostanol and furostanol Δ^5-steroid saponins

tandem mass spectroscopy (APCI-IT-MS/MS) for analysis of these 13 steroid saponins and detected the diagnostic fragment ions for the spirostanol and furostanol Δ^5-steroid saponins (Fig. 3).

The utility of CID, fast atom bombardment (FAB), electrospray ionization mass spectrometry (ESI-MS) and tandem mass spectrometry (MS/MS) for the structure elucidation of spirostanol and furostanol saponins was discussed by Liang et al. (*126*). These techniques have been applied to structure determinations of four steroidal saponins isolated from *Asparagus cochinchinensis*. In the ESI-CID spectrum, the authors observed a characteristic fragmentation involving the loss of 144 Da (Fig. 4) arising from cleavage of the E-ring when there was no sugar chain at the C-26 position. If a glucoside group was present at the C-26 position, it was preferentially eliminated. However, all compounds produced a major ion peak at $m/z = 255$ arising from the skeletal unit (*126*) and exhibited sequential loss of sugar moieties, which helped in determining the structure of the glycoside.

Although the saccharide chain and the aglycone could thus be identified by mass spectrometry, it is as yet not possible to establish the configurations of glycosidic linkages by this technique.

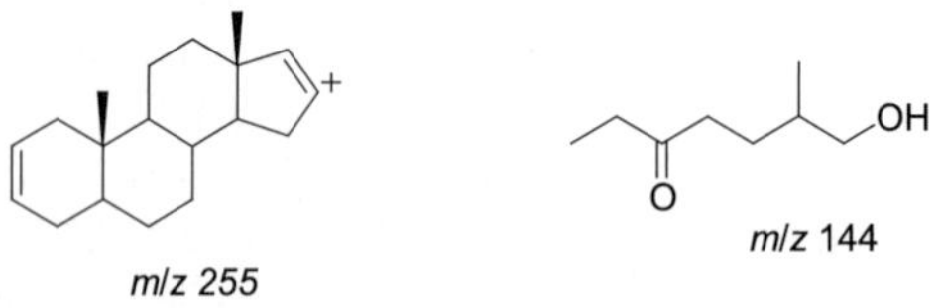

Fig. 4

References, pp. 127–141

3.3.2. NMR Spectroscopy

Of all physical methods NMR techniques have changed most during the last two decades. The introduction of high field instruments and multidimensional NMR techniques has greatly advanced structure studies of saponins. Information about the aglycone, the nature and number of the constituent sugar units including their ring sizes, anomeric configurations, interglycosidic linkages as well as the point(s) of attachment of the sugar chain to the aglycone can be obtained more readily by this method than by any other.

The first step in the structure elucidation of a saponin is to obtain the 1D ^{1}H and ^{13}C-NMR spectra. Saponins are usually investigated as deuterium exchanged samples and the most commonly used solvent is pyridine-d_5, although the use of methanol-d_4 or DMSO-d_6 has been reported in the literature. Hydroxylic protons can be exchanged by adding few drops of D_2O when required.

3.3.2.1. ^{1}H NMR Spectroscopy

The ^{1}H NMR spectra of steroid glycosides display some recognizable signals. The location of two singlets and two doublets in the region 0.5–1.7 ppm due to the methyl groups at C-10, C-13, C-20 and C-25 is very helpful in structure determination of the aglycone. The ^{1}H NMR chemical shifts of the geminal protons at C-26 assist in establishing the nature of the steroid part (spirostane or furostane) and also the stereochemistry of the methyl group at C-25. In spirostane type compounds with an equatorial 27α-methyl group, the geminal protons at C-26 resonate in a narrow range between $\delta = 3.26$–3.59 and the methyl group at C-25 resonates at $\delta = 0.57$–0.83; in contrast, in compounds with an axial 27β-methyl group, the C-26 geminal proton signals appear distinctly at $\delta = 3.18$–3.42 and 3.88–4.11 while the methyl group at C-25 resonates in the region $\delta = 0.95$–1.14 (127). In furostane steroids, a comparison of the ^{1}H NMR chemical shift data reflects several interesting characteristics. The resonances of H$_2$-26 are more resolved in the spectra of 25(S) compounds than in their 25(R) counterparts ($\Delta\delta$ is usually >0.57 ppm in 25S compounds and <0.48 ppm in 25R compounds). The signals appear in the ranges $\delta = 3.42$–3.52 and 4.02–4.10 ppm, respectively, in 25S compounds but at $\delta = 3.52$–3.63 and 3.92–3.98 ppm, respectively, in 25R isomers. The methyl group at C-25 resonates at $\delta = 0.97$–1.10 in the S isomers but somewhat upfield, at $\delta = 0.92$–1.03, in the R isomers (128).

The ^{1}H NMR spectrum also provides information about the location of the double bonds. The olefinic hydrogen at C-6 and the exomethylene protons of C-27 in spirostane analogues resonate at $\sim\delta = 5.26$–5.53 (*129*) and 4.80–4.83 (*130*), respectively, while 23-H in Δ^{22} furostane analogues resonates at $\delta = 4.60$ (*131*). Although most of the sugar protons resonate in a narrow range ($\delta = 3.0$–4.5) leading to much overlap, at least the anomeric protons are clearly distinguishable. Their signals are usually found as doublets with coupling constants 6.5–9.0 or 1.5–4.0 Hz in the region $\delta = 4.1$–6.4 ppm (*132*). Methyl doublets ($J = 6$ Hz) of 6-deoxy sugar units appear at $\delta = 1.3$–1.5 (*35, 133*).

3.3.2.2. ^{13}C NMR Spectroscopy

^{13}C NMR spectroscopy has played an important role in structure elucidation of steroidal glycosides. The spectra give a better dispersion over a 200 ppm range and the protonation levels are deducible from a DEPT experiment (*134*). Resonances of the sugar anomeric carbons are found in the well separated chemical shift range of $\delta = 96$–112 ppm, while those of the non-anomeric carbons are in the range $\delta = 60$–90 ppm, which provides information about the number of monosaccharide units present and sometimes also about the nature of the glycosidic linkages. The C-1 signals of β-anomers usually appear 2–6 ppm downfield from their α-counterparts (*132*). Glycosylation causes a downfield shift of 7–12 ppm for the α-carbon and an upfield shift of 2–5 ppm for the β-carbon (*35*). Methyl groups attached to C-10, C-13, C-20, and C-25 resonate in the region $\delta = 14$–24, 14–17, 12–17, and 16–18 ppm, respectively. Variation in the stereochemistry of the ring junction affects the chemical shifts of the angular methyl groups as well as those of other neighbouring carbons. Significant differences in the resonance positions of several carbons within rings A and B have been reported for 5α-H and 5β-H steroids (*viz.* tigogenin and smilagenin). Thus, chemical shifts for C-3, C-4, C-5, C-6, C-7, and C-19 of tigogenin are $\delta = 77.9$, 35.0, 44.8, 29.1, 32.6 and 12.5, respectively (*135*), while those for smilagenin are 75.0, 30.0, 36.0, 26.6, 26.4 and 23.7, respectively (*38*).

When a double bond in a spirostane is located at Δ^5 or $\Delta^{25(27)}$, the involved carbons resonate at $\sim\delta = 138.0$ (C-5), 125.1 (C-6), 144.5 (C-25), and 108.6 (C-27) (*129*). In the case of furostane analogues with a double bond located at 20(22) or 22(23), the carbon signals appeared at ~ 103.5 (C-20), 152.4 (C-22) or at 157.4 (C-22), 96.2 (C-23) (*131, 136, 137*).

^{13}C NMR spectrometry is also very helpful in assignment of stereochemistry at C-25 (*R/S*) of spirostane type steroidal saponins and sapogenins. Agarwal *et al.* (*138*) have studied in detail the carbon resonances of smilagenin and sarsasapogenin using DEPT, COSY, TOCSY, HETCOR,

References, pp. 127–141

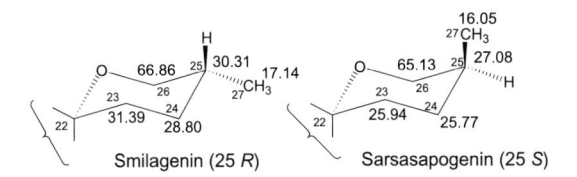

Fig. 5

HMQC, HMQC-TOCSY, HSQC-RELAY, HMBC and selective reverse INEPT techniques. As expected, the major differences (Fig. 5) were observed in the ring F resonances. All the carbon resonances except C-22 ($\sim \delta = 109.0$) occur at higher field in sarsapogenin than in smilagenin [shift differences ($\delta_2 - \delta_1$): C-22 (0.48), C-23 (5.45), C-25 (3.23), C-26 (1.73), and C-27 (1.09)].

3.3.2.3. 2D NMR Spectroscopy

The identity of the aglycone, the sugars, and the sugar sequence of the oligosaccharide chain can be determined by a combination of 2D NMR techniques like COSY (*139*), HOHAHA (*140, 141*) or TOCSY (*142*), HETCOR (*143*) or HMQC (*144*), HMBC (*145*), and NOESY (*146, 147*) or ROESY (*148, 149*). The DQF-COSY or HSQC-TOCSY spectra generally identify the fragments (short spin systems); these are linked to each other using the information obtained from NOESY/ROESY and HMBC. Careful analysis of the ^{1}H and ^{13}C NMR spectra then suggests whether tracing along the ^{1}H, ^{1}H coupling network (DQF-COSY, TOCSY or HSQC-TOCSY) will be enough or whether HMBC/INADEQUATE experiments (where proton density is low) are required for determining the structure. To establish the structure of the steroid nucleus, HMBC correlations from the angular methyl groups (18-CH$_3$, 19-CH$_3$) are most helpful. Commonly the 18-CH$_3$ proton signals display correlations with C-12, C-13, C-14, and C-17, whereas the 19-CH$_3$ signals show correlations with C-1, C-5, C-9, and C-10. From the results of the HMBC spectra and the fragments obtained from the COSY, HETCOR/HSQC and TOCSY spectra, it is possible to construct the steroid skeleton and identify the functional groups too. Furthermore the key correlations observed in the NOESY/ROESY spectra help to establish the configurations of the ring junctions. A few key NOEs will help to quickly establish the configurations at the ring junctions; thus NOESY correlations between H-1β, H-11α, and H-7β, H-15α generally indicate the *trans* fusion of the rings A and B, and rings C and D in steroid skeleton. The NMR analysis of steroids and natural products has been recently reviewed by Croasmun and Carlson (*150*), as well as by Bross-Walch *et al.* (*151*).

The identity of the sugars and the sequence of the oligosaccharide chain can also be established by a combination of 2D NMR techniques. Since the anomeric protons of each sugar residue resonate in a characteristic region well isolated from those of the other sugar protons, they are the preferred starting points for analyzing the spectra. Although a COSY spectrum, preferentially DQF-COSY (*152*), may sequentially identify all the proton signals of a monosaccharide unit starting from the anomeric proton resonance, some ambiguity may result due to signal overlap. The easiest course is to take the help of a HOHAHA/TOCSY spectrum, which optimally detects protons 3 to 5 bonds away. Sometimes, several HOHAHA experiments (*153*) with different mixing times may be necessary to trace the spin systems from the anomeric to the terminal proton step by step. Once the ^{1}H resonances have been completely assigned, ^{13}C signals can be assigned unambiguously with the help of a HETCOR or HMQC experiment. Moreover sugar residues can also be identified by comparing the ^{13}C chemical shifts with those of standard methyl glycosides or from the available literature data on steroidal saponins. The anomeric configuration can then be deduced from the magnitude of the $^3J_{H,H}$ coupling between H-1 and H-2 (large, $\sim$7–9 Hz, for diaxial orientation but much smaller, $\sim$1–3 Hz, for axial/equatorial or diequatorial arrangement) and by comparing the chemical shift of the anomeric carbon with published data. The difference in $^1J_{C1,H1}$ coupling constants between the α- and β-isomers of sugars also indicates their anomeric configurations ($^4C_1$ or $^1C_4$); the values are 167–170 Hz for the α-anomers but 158–160 Hz for β-anomers (*133, 154, 155*).

After identification of each sugar residue and the anomeric configuration, all that is required is to identify the sugar sequence and the interglycosidic linkage. It is necessary to make use of either homonuclear dipolar coupling (NOE measurements) or the long range hetero nuclear coupling constant $^3J_{CH}$ across the glycosidic linkages. The presence of an inter-glycosidic NOE from the anomeric proton of a particular sugar residue to a proton of the other sugar or non-sugar residue (sapogenin) defines the glycosidic linkage between the two residues. NOE connectivities are most often observed between the anomeric proton and the proton connected to the carbon atom of the linkage. This has been found to be of wide applicability in structure determination of naturally occurring glycosides. The conventional NOEs can be positive or negative and pass through zero when $\omega_0\tau_c$, the product of spectrometer angular frequency and molecular rotational correlation time that depends on the size and shape of the molecules and on the viscosity of the rotating medium, is approximately equal to unity. The problem, which is typical of middle sized molecules like glycosides, can be solved by performing the experiment in the rotating frame, the so-called ROESY (*156*, *157*). An example of a ROESY spectrum is shown in Fig. 6 illustrating the structure study of racemoside A (*35*).

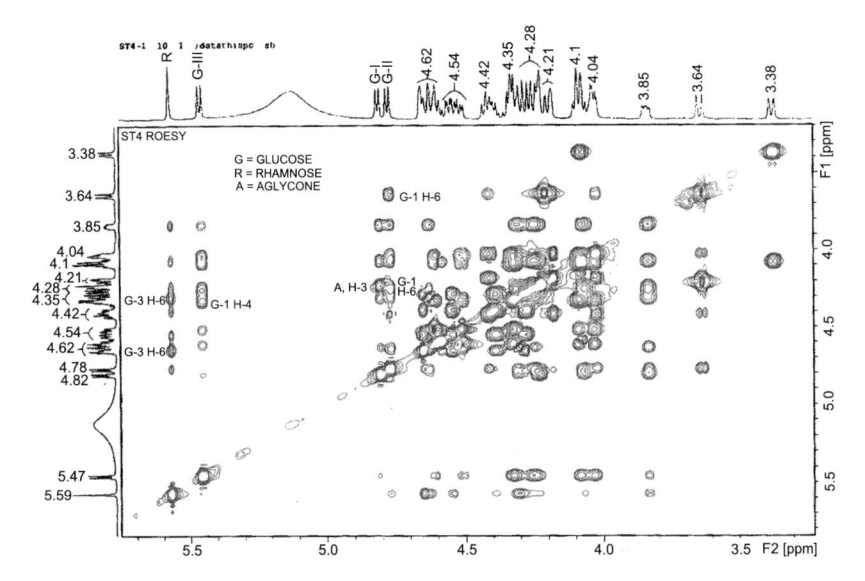

Fig. 6. ROESY spectrum of racemoside A from *Asparagus racemosus*

Correlations observed in the ROESY spectrum of racemoside A establishing the sugar-sugar and sugar-aglycone linkage

However, in rare cases the observed NOEs may be inconclusive if the chemical shift of the aglyconic proton located at the glycosylated carbon coincides with the chemical shifts for protons of other sugar moieties. This usually happens in the case of complex saponins. Therefore, NOEs should not be used as the sole criterion for establishing the position of a glycosidic linkage, especially when dealing with branching centers of the oligosaccharide chain, *e.g.* the saponin mimusopin from the seeds of *Mimusops elengi (133)*.

A more effective way to determine the sugar linkage and sequence is to detect the long-range $^3J_{CH}$ coupling across the glycosidic bond. The most practical technique is heteronuclear multibond correlation (HMBC). An HMBC experiment can furnish multi-bond correlation between the anomeric proton and the aglycone carbon or sugar carbon to which it is linked and thus serve to identify the linkage. The three bond carbon-proton couplings also follow the Karplus relationship, the maximum being usually observed at a dihedral angle of 180° and the minimum near about 90°. So, HMBC also furnishes information regarding anomeric configurations *(158)*.

4. Biological Activity

Saponins have varied biological properties that have attracted the attention of mankind since ancient times. Although they are highly toxic when given intravenously to higher animals, their toxicity is much less when administered orally *(159)*. They are more water-soluble than their aglycones as the attachment of a carbohydrate chain to the aglycone moiety increases hydrophilicity, which influences the pharmacokinetic properties of the compounds in circulation, concentration in the body fluids and elimination. Moreover, some glucosides can be transported as such into brain tissue using the glucose-transport system. Furthermore, the ability of saponins to form pores in membranes has contributed to their common use in physiological research *(160–162)*. Earlier studies on the bioactivity of saponins were conducted mainly with crude saponin mixtures containing not only saponins but also other constituents present in the extract. The advent of modern sophisticated techniques of isolation and structure determination prompted many researchers to study the biological activity of homogeneous saponins or fractions containing only saponins. In recent years there have been several reviews dealing with biological activity of saponins *(163–168)*. In the following section, information on biological activities of steroid saponins reported during the period 1999 to mid 2006 is given.

References, pp. 127–141

4.1. Cytotoxic Activity Against Cancer Cell Lines

Mimaki et al. (169) have isolated eighteen steroidal saponins from the rhizomes of Hosta sieboldii and evaluated their cytotoxic activity against human promyelocytic leukemia HL-60 cells following a modified method of Sargent and Taylor (170). The compounds were found to be less potent compared with the standard antileukaemic drugs etoposide and methotrexate. Gitogenin diglycoside and tigogenin triglycoside exhibited cytostatic activity with IC_{50} values of 3.0 and 4.5 µg ml^{-1}, respectively, but introduction of a hydroxyl group at the C-2 position of tigogenin enhances the activity to 2.8 µg ml^{-1}. Removal of the rhamnosyl unit from gitogenin diglycoside and introduction of a hydroxyl group at C-12 of gitogenin caused the activity to fall (to more than 10 µg ml^{-1}). Furostanol saponins showed considerable activity, IC_{50} ranging from 3.0 to 5.9 µg ml^{-1}. Glycosides possessing a glucosyl-$(1 \rightarrow 2)$-glucosyl-$(1 \rightarrow 4)$-galactosyl moiety as the common saccharide sequence at the C-3 position of gitogenin inhibited cell proliferation with an IC_{50} value of 3 µg ml^{-1}. However, modification of the aglycone moiety either with a C-12 carbonyl (manogenin) or a conjugated carbonyl (9,11-dehydromanogenin) decreases the activity by half to one third or more.

Phytochemical examination of fresh bulbs of Allium jesdianum, which is native to Iran and Iraq but cultivated in Japan as a garden plant with purple-lilac flowers, yielded four steroidal glycosides that were evaluated for cytotoxic activity against HL-60 human promyelocytic leukemia cells (171). One of the compounds exhibited considerable cytotoxic activity with an IC_{50} value of 1.5 µg ml^{-1} compared with etoposide used as a positive control (IC_{50} 0.3 µg ml^{-1}), while other compounds were inactive ($IC_{50} > 10$ µg ml^{-1}). The authors concluded that introduction of a hydroxyl group at C-6 of the spirostane skeleton caused the activity to decrease, while compounds belonging to the cholestane series showed no activity. Evaluation of the active glycoside in the National Cancer Institute 60 cell line assay (172) showed that the mean concentrations required to achieve GI_{50}, TGI and LC_{50} levels against the panel of cells tested were in the order of 4.5, 18, and 54 µM, respectively. However the bioactive spirostane glycoside was also relatively active against the human T cell lymphoblast-like cell line (CCRF-CEM), non-small cell lung cancer HOP-62, and breast cancer MCF-7 cells.

Ruscogenin diglycoside (glycosylation at C-1 of the genin) with three acetyl groups attached to the inner galactosyl moiety and its corresponding 26-glucosyloxyfurostanol saponin from the underground part of Ruscus aculeatus (63) exhibited 98.2 and 82.5% inhibition at

$10\,\mu g\,ml^{-1}$, respectively, against leukemia HL-60 cells, whereas two other steroidal saponins of neoruscogenin and its corresponding furostanol glucoside from the same source showed inhibitory effects against the same cell line gave IC_{50} values of 3.0 and $3.5\,\mu g\,ml^{-1}$, respectively (65). This suggested that the acetyl and 2-hydroxy-3-methylpentanoyl groups attached to the sugar moiety contribute to the cytotoxic activity. Twelve steroidal saponins isolated from the bulbs of *Allium karataviense* (81) were evaluated for cytostatic activity against human promyelocytic leukemia HL-60 cells. Only the spirostanol (25R) glycosides exhibited cytostatic activity using etoposide as a positive control.

Three new spirostanol glycosides and a bisdesmosidic cholestane glycoside from the aerial parts of *Polianthes tuberosa* (173) were evaluated for cytotoxic activity on HL-60 human promyelocytic leukemia cells. Although the cholestane glycoside did not show any activity, the spirostanol glycosides showed moderate activity. Mimaki *et al.* isolated a number of steroidal glycosides from the aerial parts of *Dracaena draco* (174), and studied their cytotoxic activity against HL-60 cells. Diosgenin-rhamno-glucoside isolated earlier from *Trillium kamtschaticum* (175), and (23S,24S)-spirosta-5,25(27)-diene glycoside showed relatively potent cytostatic activity when compared with the standard drug etoposide.

The steroidal saponins gracillin, methyl protogracillin and methyl protoneogracillin from the rhizomes of *Dioscorea collettii* var. *hypoglauca* were evaluated for cytotoxicity against human cancer cell lines from leukemia and eight solid tumor diseases (176, 177). Methyl protoneogracillin exhibited strong cytotoxic effects against two leukemia cell lines, one colon cancer line, two CNS cancer lines, one melanoma line, one renal cancer line, one prostate cancer line, and one breast cancer line. Moderate activity was also observed against four NSCLC lines, one colon cancer line, one CNS cancer line, two melanoma lines, four ovarian cancer lines, three renal cancer lines, and four breast cancer lines. Gracillin was cytotoxic against most cell lines with GI_{50}, TGI and LC_{50} at micromolar levels, but no activity was observed against non-small cell lung cancer, colon cancer, ovarian cancer, and renal cancer. Preliminary toxicity studies indicated that the maximum tolerated dose for methyl protoneogracillin in mice was 600 mg/kg (177). Regarding structure-activity relationships, the C-25 *R/S* configuration appears to be critical for activity against solid tumor cells, but was not critical for leukemia cells. COMPARE software analysis indicated that the mechanism(s) of action involved was a novel one (178, 179).

Isoterrestrosin B from the fruits of *Tribulus terrestris* (136) exhibited cytotoxicity against SK-MEL cells while the steroidal saponins isolated

References, pp. 127–141

from the leaves of *Cestrum nocturnum* (*180*) showed considerable cytotoxicity against HSC-2 cells. Moderate cytotoxicity was observed for yayoisaponins A–C against P388 murine leukemia cells compared with dioscin (*181*). Two out of five steroidal glycosides from the rhizomes of *Tacca chantrieri* displayed considerable cytotoxicity against HL-60 leukemia cells while the other three saponins did not show any cell growth inhibitory activity even at a concentration of $10\,\mu g\,ml^{-1}$, suggesting that the structures of both the aglycone and the sugar moieties contribute to the cytotoxicity (*182*). Both spirostanol and furostanol glycosides from *Cestrum nocturnum* (*183*) showed potent cytotoxic activity against human oral squamous cell carcinoma compared with doxorubicin. *In vitro* cytotoxic studies of steroidal saponins isolated from fresh tubers of *Polianthes tuberosa* (*135*) against Hela cells was determined using the 3-(4,5-dimethylthiazol-2-yl)-2,5-diphenyl tetrazolium bromide (MTT) colorimetric assay (*170*). Compounds with a carbonyl group at C-12 of the aglycone showed stronger cytotoxicities compared with those with no carbonyl group in the aglycone. The major steroidal saponins neosibiricosides C and D (*184*) from the rhizomes of *Polygonatum sibiricum* showed moderate cytotoxic activity *in vitro* against human MCF-7 breast cancer cells. The spirostanol saponins aspaoligonins A–C from *Asparagus oligoclonos* (*185*) were evaluated against human lung carcinoma, human ovary malignant ascites, human malignant melanoma and human central nervous system carcinoma *in vitro* using the standard SRB assay (*172*), which showed significant levels of cytotoxicity. The compounds are similar in activity to carboplatin but are much less potent than adriamycin. Degalactotigonin from *Solanum nigrum* showed better cytotoxicity *in vitro* (*186*) against human liver carcinoma, human lung carcinoma, human breast carcinoma and human glioma compared with 10-hydroxycamptothecin as calculated by the LOGIT method (*187*). The corresponding 23-*O*-glucoside and the 15-OH analogues did not show any inhibitory activity, suggesting that the aglycone moiety contributed to the cytotoxicity (*188*). Ikeda *et al.* (*189*) studied the cytotoxicity of steroidal glycosides (having the frameworks of spirostane, furostane, spirosolane, and pregnane) from *Solanum nigrum* and *S. lyratum* as well as steroidal glycosides from *Allium tuberosum* against human lung cancer (*190*) and human colon cancer (*191*) cell lines. Of the 21 compounds tested, β-lycotetraosyl spirostanol without an additional oxygen functional group in the steroid nucleus was the most effective against both cell lines. The β-lycotetraosyl derivatives of spirostanes were more cytotoxic than the chacotriose derivatives, while protodioscin and the β-lycotetraosyl derivatives of furostane glycosides proved as potent as dioscin. The activity of the compounds against human lung cancer cell line was lower when the terminal xylopyranosyl moiety was replaced with a glucopyranose unit. On comparison of the aglycone moieties it was found that glycosides having 25S stereochemistry showed almost no activity, whereas those with 25R stereochemistry were as active as the standard drug CDDP, suggesting that the C-25 position might play an important role in mediating cytotoxicity. Furthermore the presence of oxygenated functional groups on the aglycones reduced the activity.

Hernández et al. (*192*) reported that icogenin, a furostanol glycoside, inhibited the growth and viability of HL-60 cells in a dose dependent manner as determined by the MTT dye-reduction assay method (*193*). Growth inhibition was caused by induction of apoptosis, as determined using quantitative fluorescent microscopy on nuclear changes. Furthermore, it was demonstrated by western blot analysis that the 116 kDa active poly(ADP-ribose) polymerase-1 protein was cleaved into its characteristic 85 kDa fragment after treatment of the cells with icogenin thus confirming *in vivo* activation of caspase, the main protease responsible for poly(ADP-ribose) polymerase cleavage (*194, 195*). In order to study structure activity relationships, three other spirostanol glycosides, – an acetyl derivative of a spirostanol glycoside, diosgenone and diosgenin – were also taken into account. It was found that the spirostanol or furostanol ring or the acetyl groups in the sugar moiety do not play any crucial role in cytotoxicity, but an α-L-rhamnosyl moiety attached to C-2 of the inner glucosyl moiety has more substantial effects.

Steroidal saponins have been found to have potent antiproliferative activity. The saponins from the roots and rhizomes of *Dracaena angustifolia* (*196*) were tested for antiproliferative activity against human HT-1080 fibrosarcoma, murine colon 26-L5 carcinoma and B-16 BL6 melanoma cell lines. Cellular viability in the presence or absence of test samples was determined following the standard assay method (*197*). The results indicated that the spirostanol saponins possess a greater antiproliferative activity compared with their furostanol analogues. A 24-*O*-fucopyranosyl unit and a xylopranosyl unit in the inner glucose moiety attached to C-3 of the aglycone seem to be important for cytotoxic activity against HT-1080 fibrosarcoma cells. The IC_{50} values varied from 0.2 to 3.8 µM compared to 0.2 µM for the positive control doxorubicin.

4.2. Antifungal Activity

The antifungal activity of steroidal saponins against agricultural pathogens has been known for a long time (*198–201*) and several patents have been issued (*202–205*). Many steroidal saponins exhibit antifungal activity under experimental conditions. Yang *et al.* studied the antifungal activity of 22 steroidal saponins and six steroidal sapogenins isolated from a number of monocotyledons against *Candida albicans, C. glabrata, C. krusei, Cryptococcus neoformans* and *Aspergillus niger.* The aglycone moieties of the steroidal saponins were hecogenin, neohecogenin, tigogenin, neotigogenin, chlorogenin, or diosgenin. Four saponins with tigogenin as aglycone and a sugar moiety of four or five monosaccharide units exhibited significant activity against *C. neofoemans* and *A. fumigatus* comparable to the positive control amphotericin B, suggesting that the C_{27}-steroidal saponins may be considered as potential antifungal agents (*206*).

The antifungal activity of eight steroidal saponins isolated from *Smilacina atropurpurea* (*207*) was tested following a modified version of the NCCLS methods (*208, 209*). Among them two, atropurosides B and F, were found to be moderately active against *Candida albicans, C. glabrata, Cryptococcus neoformans* and *Aspergillus fumigatus*, while dioscin, one of the major components of the plant, was more active than amphotericin B against *C. albicans* and *C. glabrata*. Antifungal activity *in vitro* was also detected in the crude extract from *Yucca gloriosa* against *Candida albicans, C. tropicalis, C. glabrata, C. krusei* and *C. kefyr.* The two spirostanol glycosides yuccaloesides B and C isolated from the plant exhibited fungicidal activity and were as effective as amphotericin B and ketoconazole (*210*). The results are quite close to those reported by Miyakoshi *et al.* in a study of steroidal saponins from *Y. schidigera* (*211*) used as an antideteriorating agent in foods. The saponins with a branched-chain trisaccharide unit without any oxygen functionalities at C-2 and C-12 exhibited potent antiyeast activities, while saponins with a 2β-hydroxyl or 12-keto group showed very weak or no activity.

The antifungal activities of the steroidal saponins isolated from *Solanum hispidum* and *S. chrysotrichum* possessing 25*S* and 25*R* stereochemistry were studied following the conventional agar dilution assay procedure (*212*). Spirostanol glycosides with 25*R* configuration and a disaccharide moiety [xylose (1 → 3) quinovose] at C-6 of the aglycone were shown to exhibit a broad spectrum of activity against yeast as well as dermatophyte species. The structure activity relationships were discussed (*213, 214*). Steroidal saponins isolated from *Smilax medica* were evaluated for antifungal activity against the human pathogenic yeasts *Candida albicans, C. glabrata* and *C. tropicalis*. Compounds having a spirostane skeleton exhibited antifungal activity against the three yeasts tested, while the compound with a furostane skeleton showed negative results, suggesting that the E and F rings of spirostane-type steroids play a key role in the mediation of antifungal properties (*38*). These results were also in agreement with earlier publications (*215–217*).

4.3. Miscellaneous Effects

The hemolytic properties of steroidal saponins isolated from *Agave* species have been investigated and reviewed (*218*). A steroidal saponin isolated from *A. attenuata* was shown to possess powerful hemolytic properties (*219*) when compared with adjuvants commonly used in animal and human experimental models by an *in vitro* assay method (*220*).

Aphids are sap-feeding insects causing direct damage to the agricultural crops and are virus vectors (*221, 222*). Luciamin, isolated from *Solanum laxum*, exerts a deterrent effect on aphids and was the first steroidal glycoside found to possess this property (*223*).

The hypocholesterolaemic effects of several saponins in a variety of experimental animals have been reported (*224*). Koch (*225*) indicated that the cholesterol-lowering effect of garlic preparations may be due to its saponin content. Cholesterol-lowering effects of the saponin fractions from garlic rich in steroidal saponins have been studied in rat models. Plasma total and LDL cholesterol concentration levels decreased significantly without change of HDL cholesterol levels in all rat groups when they were fed with 0.3 g/kg/day garlic extract for 16 weeks. The author has suggested that special consideration should be given to steroid saponins besides organosulphur compounds in biological and pharmacological studies of garlic and its preparations (*226*).

Torvanol A and torvoside H isolated from *Solanum torvum* (*76*) showed antiviral activity (herpes simplex virus type 1) *in vitro*. The IC_{50} values were threefold less compared with the reference compound, acyclovir.

Leishmaniasis is a public health problem throughout most of the tropical and subtropical world, and the visceral form is the most fatal if left untreated. To date, there are no vaccines against visceral leishmaniasis and chemotherapy is the main weapon in the physician's arsenal. The first line of treatment is losing its effectiveness due to parasite resistance while others are toxic, expensive and prone to resistance development. Racemoside A, a steroidal saponin isolated from *Asparagus racemosus*, is a potent anti-leishmanial agent effective (*in vitro*) against antimony sensitive (AG83, $IC_{50} = 1.25\,\mu g/ml$) as well as unresponsive (GE1F8R, $IC_{50} = 1.61\,\mu g/ml$) *L. donovani* promastigotes, and exerts its leishmanicidal effect through induction of programmed cell death. Racemoside A caused plasma membrane alteration as measured by Annexin V and PI binding, loss of mitochondrial membrane potential culminating in cell cycle arrest at sub G0/G1 phase, and DNA nicking as evidenced from deoxynucleotidyltransferase-mediated dUTP end labeling (TUNEL). Morphological alterations include cell shrinkage, aflagellated ovoid shape and chromatin condensation. The compound is also effective against amastigotes (*ex vivo*) of *L. donovani* (AG83, $IC_{50} = 0.17\,\mu g/ml$) but is almost nontoxic to the murine peritoneal macrophages even up to a higher concentration of $10\,\mu g/ml$ (viability >89%). Racemoside A can be considered as a potent antileishmanial agent meriting further pharmacological investigation (*35, 227*).

5. Biosynthesis of Steroidal Glycosides

Plants synthesize diverse classes of secondary metabolites, including steroidal saponins, mainly to defend themselves against pathogen attack and pests (*228–231*). Biosynthesis of cardenolides, bufadienolides and steroidal sapogenins has been reviewed earlier by Tschesche (*15, 232*). It has been well established that the classical mevalonate pathway is involved in the synthesis of isopentenyl pyrophosphate which subsequently synthesizes the hydrocarbon squalene. The enzyme squalene monooxygenase oxidizes squalene to 2,3-oxidosqualene, the precursors of steroid sapogenins, *via* cycloartenol and cholesterol. Oxidation of cholesterol at C-16, C-22 and C-26/27, and subsequent cyclization of the oxygenated cholesterol leads to the formation of the spiroketal ring (*233*). Glucosylation of the hydroxyl group at C-26/27 takes place earlier than the formation of the spiroketal ring (*234, 235*), thus forming the furostanol 26-β-D-glucoside. The resulting furostanol glucoside would then be glycosylated effectively by the enzyme UDPGlc (*236, 237*) at the C-3 hydroxyl group to form the bisdesmosidic furostane saponins. Enzymatic removal of the C-26 glucose moiety and spontaneous cyclization to form the heterocyclic ring leads to formation of spirostane glycosides. However, it is worth mentioning that the biogenetic relationship between the furostane and spirostane derivatives is still controversial as experiments indicated that the glucosyltransferase (Gtase) from asparagus fern efficiently glucosylated the spirostane derivative yamogenin but was unable to glucosylate its furostane analogue (*238*). This proposition is supported by results obtained using cell suspension cultures of crape ginger (*239*). During the last two decades substantial progress has been made in the identification and biochemical characterization of Gtases involved in the biosynthesis of saponins and glycoalkaloids. A number of enzymes taking part in the formation or the rearrangements of the carbohydrate moieties found in these compounds have been isolated from various plant species and thoroughly characterized (*240*). However, a detailed study of gene function may be necessary to unravel the reactions taking place in the formation of such secondary plant metabolites.

6. Report of New Steroidal Saponins (1998–Mid-2006)

New steroidal saponins isolated during the period 1998–mid-2006 along with their natural distribution, available physical data and spectral data are listed in Table 1. Structures **1–173** are sapogenins of the various saponins presented in Table 1.

References, pp. 127–141

Table 1. Steroidal saponins isolated during 1998–mid-2006

Plant name and family	Glycoside, physical nature, mp (°C), Mol. formula, Mol. wt. (m/z), $[\alpha]_D$	Aglycone/sapogenin	Sugar with linkage	Reference
Agave americana (Agavaceae)	Agamenoside H, AP, $C_{39}H_{64}O_{16}$, HR-FAB-MS: 787.4177 $[M-H]^-$, $[\alpha]_D^{21}$ −42.1° (c 0.011, Pyr)	Agavegenin C (**29**)	-6-O-β-D-Glup; -24-O-β-D-Glup	*241*
	Agamenoside I, AP, $C_{33}H_{54}O_{10}$, HR-FAB-MS: 609.3676 $[M-H]^-$, $[\alpha]_D^{14}$ −39.9° (c 0.041, Pyr)	(22S, 23S, 24R, 25S)-5α-Spirostane-3β,23,24-triol (**28**)	-24-O-β-D-Glup	
	Agamenoside J, AP, $C_{33}H_{54}O_{10}$, HR-FAB-MS: 609.3590 $[M-H]^-$, $[\alpha]_D^{21}$ −37.1° (c 0.018, Pyr)	(22S, 23S, 25R, 26S)-23,26-Epoxy-5α-furostane-3β,22,26-triol (**152**)	-26-O-β-D-Glup	
A. attenuata	Compound 1, colorless needles, 245–250°C, LSI-MS: 1225 $[M-H]^-$, $[\alpha]_D^{25}$ −220.0° (c 1.0, MeOH)	Sarsasapogenin (**34**)	-3-O-β-D-Glup-$(1 \rightarrow 2)$-β-D-Glup-$(1 \rightarrow 2)$-O-[β-D-Glup-$(1 \rightarrow 3)$]-β-D-Glup-$(1 \rightarrow 4)$-β-D-Galp	220
A. attenuata	Compound 1, colorless needles, 225–235°C, $C_{64}H_{108}O_{34}$, LSI-MS: 1419 $[M-H]^-$, $[\alpha]_D^{25}$ −280.0° (c 1.0, MeOH)	(25S)-22α-Methoxy-5β-furostane-3β,26-diol (**111**)	-3-O-β-D-Glup-$(1 \rightarrow 2)$-β-D-Glup-$(1 \rightarrow 2)$-O-[β-D-Glup-$(1 \rightarrow 3)$]-β-D-Glup-$(1 \rightarrow 4)$-β-D-Galp	242

Table 1 (continued)

Plant name and family	Glycoside, physical nature, mp (°C), Mol. formula, Mol. wt. (m/z), $[\alpha]_D$	Aglycone/sapogenin	Sugar with linkage	Reference
A. brittoniana	Compound 1	(25R)-5α-Spirostane-3β,6α-diol-12-one (**10**)	3-{(O-6-deoxy-α-L-Manp-(1→4)-O-β-D-Glup-(1→3)-O-[O-β-D-Glup-(1→3)-β-D-Glup-(1→2)]-O-β-D-Glup-(1→4)-β-D-Galp}	243
A. decipiens	Saponin-I, WP, 271–272°C, CI-MS: 1063 [M–H]$^-$	(25R)-22α-Methoxy-furost-5-ene-3β,26-diol (**131**)	3-O-α-L-Rhap-(1→2)-α-L-Rhap-(1→4)-β-D-Glup; -26-O-β-D-Glup	244
	Saponin-II, WP, 255–257°C, CI-MS: 1079 [M–H]$^-$	Neoruscogenin (**85**)	1-O-β-D-Glup-(1→3)-α-L-Rhap-(1→2)-β-D-Glup-(1→4)-β-D-Galp	
	Saponin-III, WP, 258–260°C, CI-MS: 1077 [M–H]$^-$	22ξ-Methoxy-furosta-5,25(27)-diene-1β,3β,26-triol (**123**)	1-O-α-L-Rhap-(1→2)-α-L-Rhap-(1→4)-β-D-Glup; -26-O-β-D-Glup	

Steroidal Saponins

Species	Compound	Aglycone	Sugar chain	Ref.
	Saponin-IV, WP, 248–250°C, CI-MS: 1211.9 [M−H]⁻	Neohecogenin (**31**)	-3-O-β-D-Glup- (1 → 3)-β-D-Xylp- (1 → 3)-β-D-Xylp- (1 → 2)-β-D-Glup- (1 → 4)-β-D-Galp	
A. fourcroydes	Compound 1, WAS, C₆₃H₁₀₄O₃₃, HR-FAB-MS: 1411.6333 [M+Na]⁺, [α]$_D^{24}$ −16.6° (c 0.78, Pyr)	β-Chlorogenin (**11**)	-3-O-[α-L-Rhap- (1 → 4)-β-D-Glup- (1 → 3)-{β-D-Glup- (1 → 3)-β-D-Glup- (1 → 2)}-β-D-Glup- (1 → 4)-β-D-Galp]	36
A. shrevei	Compound 1	(25R)-22ξ-Methoxy-5α-furostane-3β, 26-diol (**93**)	-3-O-β-D-Glup- (1 → 2)-O-[O-β-D-Glup-(1 → 4)-O-[O-β-D-Glup-(1 → 6)]-O-β-D-Glup-(1 → 4)-β-D-Galp; -26-O-β-D-Glup	245
Allium ampleoprasum (Liliaceae)	Yayoisaponin A, AS, C₅₆H₉₁O₂₉, HR-FAB-MS: 1227.5670 [M−H]⁻, [α]$_D^{23}$ −44.5° (c 0.50, Pyr)	Agigenin (**20**)	-3-O-β-D-Glup- (1 → 3)-β-D-Glup- (1 → 2)-[β-D-Xylp- (1 → 3)]-β-D-Glup- (1 → 4)-β-D-Galp	181
	Yayoisaponin B, AS, C₅₆H₈₉O₂₉, HR-FAB-MS: 1225.5507 [M−H]⁻, [α]$_D^{23}$ −23.0° (c 0.02, Pyr)	Porrigenin B (**12**)	-3-O-β-D-Glup- (1 → 3)-β-D-Glup- (1 → 2)-[β-D-Xylp- (1 → 3)]-β-D-Glup- (1 → 4)-β-D-Galp	

Table 1 (continued)

Plant name and family	Glycoside, physical nature, mp (°C), Mol. formula, Mol. wt. (m/z), $[\alpha]_D$	Aglycone/sapogenin	Sugar with linkage	Reference
	Yayoisaponin C, AS, $C_{51}H_{83}O_{25}$, HR-FAB-MS: 1095.5254 [M−H]$^-$, $[\alpha]_D^{23}$ −41.4° (c 0.21, Pyr)	Agigenin (**20**)	-3-O-β-D-Glup-$(1 \rightarrow 2)$-[β-D-Glup-$(1 \rightarrow 3)$]-β-D-Glup-$(1 \rightarrow 4)$-β-D-Galp	
A. elburzense (Alliaceae)	Elburzenoside A1, AS, $C_{39}H_{66}O_{17}$, HR-FAB-MS: 806.9276 [M−H]$^-$, $[\alpha]_D^{25}$ −41.67° (c 0.1, MeOH)	Furostane-2α,3β,5α, 6β,22α,26-hexol (**109**)	-3-O-β-D-Glup; -26-O-β-D-Glup	246
	Elburzenoside A2, AS, $C_{39}H_{66}O_{17}$, HR-FAB-MS: 806.9278 [M−H]$^-$, $[\alpha]_D^{25}$ −41.65° (c 0.1, MeOH)	Furostane-2α,3β,5α, 6β,22β,26-hexol (**110**)	-3-O-β-D-Glup; -26-O-β-D-Glup	
	Elburzenoside B1, AS, $C_{45}H_{76}O_{22}$, HR-FAB-MS: 969.0675 [M−H]$^-$, $[\alpha]_D^{25}$ −43.59° (c 0.1, MeOH)	Furostane-2α,3β,5α, 6β,22α,26-hexol (**109**)	-3-O-[β-D-Glup-$(1 \rightarrow 4)$-O-β-D-Glup]; -26-O-β-D-Glup	
	Elburzenoside B2, AS, $C_{45}H_{76}O_{22}$, HR-FAB-MS: 969.0679 [M−H]$^-$, $[\alpha]_D^{25}$ −43.61° (c 0.1, MeOH)	Furostane-2α,3β,5α, 6β,22β,26-hexol (**110**)	-3-O-[β-D-Glup-$(1 \rightarrow 4)$-O-β-D-Glup]; -26-O-β-D-Glup	
	Elburzenoside C1, AS, $C_{39}H_{66}O_{16}$, HR-FAB-MS: 790.9285 [M−H]$^-$, $[\alpha]_D^{25}$ −13.72° (c 0.1, MeOH)	Furostane-2α,3β,5α, 22α,26-pentol (**106**)	-3-O-β-D-Glup; -26-O-β-D-Glup	
	Elburzenoside C2, AS, $C_{39}H_{66}O_{16}$, HR-FAB-MS: 790.9280 [M−H]$^-$, $[\alpha]_D^{25}$ −13.70° (c 0.1, MeOH)	Furostane-2α,3β,5α, 22β,26-pentol (**107**)	-3-O-β-D-Glup; -26-O-β-D-Glup	

	Elburzenoside D1, AS, $C_{50}H_{84}O_{25}$, HR-FAB-MS: 1084.9477 [M−H]$^-$, $[\alpha]_D^{25}$ −23.75° (c 0.1, MeOH)	Furostane-$2\alpha,3\beta,5\alpha,22\alpha,26$-pentol (**106**)	-3-O-[β-D-Xylp-$(1 \to 3)$-O-β-D-Glup-$(1 \to 4)$-O-β-D-Galp]; -26-O-β-D-Glup	
	Elburzenoside D2, AS, $C_{50}H_{84}O_{25}$, HR-FAB-MS: 1084.9480 [M−H]$^-$, $[\alpha]_D^{25}$ −23.65° (c 0.1, MeOH)	Furostane-$2\alpha,3\beta,5\alpha,22\beta,26$-pentol (**107**)	-3-O-[β-D-Xylp-$(1 \to 3)$-O-β-D-Glup-$(1 \to 4)$-O-β-D-Galp]; -26-O-β-D-Glup	
A. jesdianum	Compound 4, AS, $C_{50}H_{82}O_{24}$, HR-FAB-MS: 1089.5111 [M+Na]$^+$, $[\alpha]_D^{27}$ −42.0° (c 0.1, MeOH)	$(25R)$-5α-Spirostane-$2\alpha,3\beta,6\alpha$-triol (**19**)	-3-O-{O-β-D-Glup-$(1 \to 2)$-O-[β-D-Xylp-$(1 \to 3)$]-O-β-D-Glup-$(1 \to 4)$-O-β-D-Galp}	171
A. karataviense (Liliaceae)	Compound 7, AS, $C_{37}H_{60}O_{13}$, HR-FAB-MS: 735.3959 [M+Na]$^+$, $[\alpha]_D^{27}$ −92.0° (c 0.1, MeOH)	$(25R)$-3-O-(2-Hydroxybutyryl)-5α-spirostane-$2\alpha,3\beta,5,6\beta$-tetrol (**161**)	-2-O-β-D-Glup	81
	Compound 8, AS, $C_{40}H_{58}O_{13}$, HR-FAB-MS: 747.3400 [M+H]$^+$, $[\alpha]_D^{27}$ −106.0° (c 0.1, MeOH)	$(24S, 25S)$-3-O-Benzoyl-5α-spirostane-$2\alpha,3\beta,5,6\beta,24$-pentol (**162**)	-2-O-β-D-Glup	
	Compound 9, AS, $C_{39}H_{64}O_{17}$, HR-FAB-MS: 827.4111 [M+Na]$^+$, $[\alpha]_D^{27}$ −78.0° (c 0.1, MeOH)	$(24S, 25S)$-5α-Spirostane-$2\alpha,3\beta,5,6\beta,24$-pentol (**30**)	-2-O-β-D-Glup; -24-O-β-D-Glup	
	Compound 10, AS, $C_{46}H_{68}O_{18}$, HR-FAB-MS: 931.4286 [M+Na]$^+$, $[\alpha]_D^{27}$ −60.0° (c 0.1, MeOH)	$(24S, 25S)$-3-O-Benzoyl-5α-spirostane-$2\alpha,3\beta,5,6\beta,24$-pentol (**162**)	-2-O-β-D-Glup; -24-O-β-D-Glup	
	Compound 11, AS, $C_{45}H_{74}O_{22}$, FAB-MS: 965 [M−H]$^-$, $[\alpha]_D^{27}$ −67.0° (c 0.1, MeOH)	$(24S, 25S)$-5α-Spirostane-$2\alpha,3\beta,5,6\beta,24$-pentol (**30**)	-2-O-β-D-Glup; -24-O-{O-β-D-Glup-$(1 \to 2)$-O-β-D-Glup}	

Table 1 (continued)

Plant name and family	Glycoside, physical nature, mp (°C), Mol. formula, Mol. wt. (m/z), $[\alpha]_D$	Aglycone/sapogenin	Sugar with linkage	Reference
	Compound 12, AS, $C_{40}H_{68}O_{17}$, HR-FAB-MS: 843.4301 $[M+Na]^+$, $[\alpha]_D^{27}$ −70.0° (c 0.1, MeOH)	(25R)-22ξ-Methoxy-5α-furostane-2α,3β,5,6β,26-pentol (**108**)	-2-O-β-D-Glup; -26-O-β-D-Glup	247
A. nutans (Alliaceae)	Compound 2, plates, $C_{42}H_{76}O_{19}$, LSI-MS: 883 $[M-H]^-$	Diosgenin (**49**)	-3-O-α-L-Rhap-$(1\rightarrow 2)$-[β-D-Glup-$(1\rightarrow 4)$]-O-β-D-Galp	
	Compound 3, AS, $C_{33}H_{52}O_9$, LSI-MS: 591 $[M-H]^-$	Ruscogenin (**52**)	-1-O-β-D-Galp	
A. porrum (Liliaceae)	Compound 3, FAB-MS: 1049 $[M-H]^-$, $[\alpha]_D^{25}$ −57.0° (MeOH)	β-Chlorogenin (**11**)	-3-O-{O-β-D-Glup-$(1\rightarrow 2)$-O-[β-D-Xylp-$(1\rightarrow 3)$]-O-β-D-Glup-$(1\rightarrow 4)$}-β-D-Galp}	248
	Compound 4, FAB-MS: 1211 $[M-H]^-$, $[\alpha]_D^{25}$ −56.0° (MeOH)	β-Chlorogenin (**11**)	-3-O-{O-β-D-Glup-$(1\rightarrow 3)$-β-D-Glup-$(1\rightarrow 2)$-O-[β-D-Xylp-$(1\rightarrow 3)$]-O-β-D-Glup-$(1\rightarrow 4)$}-β-D-Galp}	
A. tuberosum	Tuberoside F, AS, $C_{52}H_{86}O_{23}$, ESI-MS: 1102 $[M+Na]^+$, $[\alpha]_D^{25}$ −27.8° (c 0.22 MeOH)	(20R, 25S)-20-Methoxy-5α-furost-22-ene-2α,3β,26-triol (**166**)	-3-O-α-L-Rhap-$(1\rightarrow 2)$-[α-L-Rhap-$(1\rightarrow 4)$]-β-D-Glup; -26-O-β-D-Glup	131

	Tuberoside G, AS, $C_{51}H_{84}O_{23}$, $[\alpha]_D^{17}$ −46.0° (c 0.3, MeOH)	(20R, 25S)-5α-Furost-22-ene-2α,3β,20,26-tetrol (**165**)	3-O-α-L-Rhap-$(1 \rightarrow 2)$-[α-L-Rhap-$(1 \rightarrow 4)$]-β-D-Glup; 26-O-β-D-Glup	
	Tuberoside H, AS, $C_{51}H_{84}O_{23}$, $[\alpha]_D^{25}$ −41.8° (c 0.34, MeOH)	(20S, 25S)-5α-Furost-22-ene-2α,3β,20,26-tetrol (**164**)	3-O-α-L-Rhap-$(1 \rightarrow 2)$-[α-L-Rhap-$(1 \rightarrow 4)$]-β-D-Glup; 26-O-β-D-Glup	
	Tuberoside I, AS, $C_{51}H_{84}O_{22}$, $[\alpha]_D^{25}$ −41.8° (c 0.28, MeOH)	(20S, 25S)-5α-Furost-22-ene-3β,20,26-triol (**163**)	3-O-α-L-Rhap-$(1 \rightarrow 2)$-[α-L-Rhap-$(1 \rightarrow 4)$]-β-D-Glup; 26-O-β-D-Glup	
A. tuberosum	Tuberoside, AP, $C_{45}H_{74}O_{18}$, 292–293 °C, FAB-MS: 903 $[M+H]^+$, $[\alpha]_D^{25}$ −33.0° (c 0.02, MeOH)	Crestagenin (**22**)	3-O-α-L-Rhap-$(1 \rightarrow 2)$-O-[α-L-Rhap-$(1 \rightarrow 4)$]-β-D-Glup	249
A. tuberosum	Compound 1, AP, $C_{51}H_{86}O_{22}$, HR-FAB-MS: 1073.5509 $[M+Na]^+$, $[\alpha]_D^{29}$ −45.4° (c 0.17, Pyr)	(25R)-5α-Furostane-3β,22ξ,26-triol (**100**)	3-O-β-Chacotrioside; 26-O-β-D-Glup	250
	Compound 2, AP, $C_{45}H_{76}O_{20}$, HR-FAB-MS: 959.4833 $[M+Na]^+$, $[\alpha]_D^{29}$ −53.2° (c 0.2, Pyr)	(25S)-Furostane-3β,5β,6α,22ξ,26-pentol (**116**)	3-O-α-L-Rhap-$(1 \rightarrow 4)$-β-D-Glup; 26-O-β-D-Glup	
A. vineale	Compound 13, AP, $C_{63}H_{106}O_{34}$	(25R)-5α-Furostane-3β,6β,22ξ,26-tetrol (**105**)	3-O-β-D-Glup-$(1 \rightarrow 2)$-O-[β-D-Glup-$(1 \rightarrow 3)$]-O-β-D-Glup-$(1 \rightarrow 4)$-O-[α-L-Rhap-$(1 \rightarrow 2)$]-O-β-D-Galp; 26-O-β-D-Glup	226

Plant name and family	Glycoside, physical nature, mp (°C), Mol. formula, Mol. wt. (m/z), $[\alpha]_D$	Aglycone/sapogenin	Sugar with linkage	Reference
	Compound 20, AP, $C_{57}H_{94}O_{28}$, FAB-MS: 1225 $[M-H]^-$	β-Chlorogenin (**11**)	3-O-β-D-Glup- $(1\rightarrow 2)$-O-[β-D-Glup- $(1\rightarrow 3)$]-O-β-D-Glup- $(1\rightarrow 4)$-O-[α-L-Rhap- $(1\rightarrow 2)$]-O-β-D-Galp	
Asparagus africanus (Liliaceae)	Gloriogenin, fine needles, $C_{44}H_{70}O_{18}$, 206–207.6°C, FAB-MS: 909 $[M+Na]^+$, $[\alpha]_D^{20}$ +52.0° (c 0.18, CH_2Cl_2)	Gloriogenin (**43**)	3-O-[β-D-Glup- $(1\rightarrow 2)$-[α-L-Arap- $(1\rightarrow 6)$]-β-D-Glup}	106
	Compound 2, colorless flakes, $C_{44}H_{72}O_{17}$, 266–267.3°C, FAB-MS: 896 $[M+Na]^+$, $[\alpha]_D^{20}$ +57.0° (c 0.05, CH_2Cl_2)	Smilagenin (**35**)	3-O-[β-D-Glup- $(1\rightarrow 2)$-[α-L-Arap- $(1\rightarrow 6)$]-β-D-Glup}	
	Compound 3, fine needles, $C_{46}H_{78}O_{19}$, 160.2–161.4°C, FAB-MS: 936 $[M+H]^+$, $[\alpha]_D^{20}$ −29.0° (c 0.14, MeOH)	(25R)-22α-Methoxy-5β-furostane-3β,26-diol (**112**)	3-O-β-D-Glup- $(1\rightarrow 2)$-[β-D-Glup]; 26-O-β-D-Glup	
A. cochinchinensis (Asparagaceae)	Asparacoside 1, WP, $C_{49}H_{80}O_{21}$, HR-TOF-MS: 1027.5100 $[M+Na]^+$, $[\alpha]_D^{20}$ −35.2° (c 0.57, $CHCl_3$-MeOH, 1:1)	Sarsasapogenin (**34**)	3-O-α-L-Arap- $(1\rightarrow 6)$-[α-L-Arap- $(1\rightarrow 4)$]-[β-D-Glup- $(1\rightarrow 2)$]-β-D-Glup	251
A. filicinus (Liliaceae)	Aspafilioside D, WAP $C_{49}H_{82}O_{22}$, 190–191°C, ESI-MS: 1021 $[M-H]^-$, $[\alpha]_D^{20}$ −18.0° (c 0.27, MeOH)	(25S)-5β-Furostane-3β,22,26-triol (**115**)	3-O-β-D-Xylp- $(1\rightarrow 2)$-[β-D-Xylp- $(1\rightarrow 4)$]-β-D-Glup; 26-O-β-D-Glup	252

A. officinalis	Sarsasapogenin M, WAP, C$_{39}$H$_{64}$O$_{14}$, HR-ESI-MS: 779.4187 [M+Na]$^+$, [α]$_D^{22}$ −65.46° (c 0.25, MeOH)	(25S)-Spirostane-3β, 17α-diol (**37**)	-3-O-β-D-Glup-(1→2)-O-β-D-Glup	253
	Sarsasapogenin N, WAP, C$_{45}$H$_{74}$O$_{17}$, HR-ESI-MS: 909.4822 [M+Na]$^+$, [α]$_D^{22}$ −86.22° (c 0.11, MeOH)	(25S)-Spirostane-3β, 17α-diol (**37**)	-3-O-α-L-Rhap-(1→2)-[α-L-Rhap-(1→4)]-O-β-D-Glup	
A. oligoclonos	Aspaoligonin A, WAP, C$_{39}$H$_{64}$O$_{14}$, HR-FAB-MS: 779.4223 [M+Na]$^+$, [α]$_D^{25}$ −14.29° (c 0.05, Pyr)	(25S)-Spirostane-3β, 17α-diol (**37**)	-3-O-β-D-Glup-(1→2)-β-D-Glup	185
	Aspaoligonin B, WAP, C$_{44}$H$_{72}$O$_{17}$, HR-FAB-MS: 895.4667 [M+Na]$^+$, [α]$_D^{25}$ −62.07° (c 0.03, Pyr)	(25S)-Spirostane-3β, 17α-diol (**37**)	-3-O-α-L-Rhap-(1→4)-[β-D-Xylp-(1→2)]-β-D-Glup	
A. racemosus	Racemoside A, colorless needles, 244–246°C, C$_{51}$H$_{84}$O$_{22}$, ESI-TOF-MS: 1171 [M+Na]$^+$, [α]$_D^{26}$ −34.9° (c 0.90, MeOH)	Sarsasapogenin (**34**)	-3-O-{β-D-Glup-(1→6)-[α-L-Rhap-(1→6)-β-D-Glup-(1→4)]-β-D-Glup}	35
	Racemoside B, colorless crystals, 240–242°C, C$_{45}$H$_{74}$O$_{17}$, ESI-TOF-MS: 909 [M+Na]$^+$, [α]$_D^{26}$ −41.1° (c 0.81, MeOH)	Sarsasapogenin (**34**)	-3-O-α-L-Rhap-(1→6)-β-D-Glup-(1→6)-β-D-Glup	
	Racemoside C, colorless needles, 236–238°C, C$_{45}$H$_{74}$O$_{16}$, ESI-TOF-MS: 893 [M+Na]$^+$, [α]$_D^{26}$ −55.4° (c 0.56, MeOH)	Sarsasapogenin (**34**)	-3-O-{α-L-Rhap-(1→6)-[α-L-Rhap-(1→4)]-β-D-Glup}	

Table 1 (*continued*)

Plant name and family	Glycoside, physical nature, mp (°C), Mol. formula, Mol. wt. (*m/z*), $[\alpha]_D$	Aglycone/sapogenin	Sugar with linkage	Reference
Balanites aegyptica (Zygophyllaceae)	Compound 2, $C_{57}H_{94}O_{28}$, ESI-MS: 1179 [M−MeOH + H]$^+$, $[\alpha]_D^{21}$ −1.80° (*c* 1.67, MeOH)	(20*S*, 22*R*, 25*R*)-22-Methoxy-furost-5-ene-3β,26-diol (**135**)	3-*O*-β-D-Xylp-(1→3)-β-D-Glup-(1→4)-[α-L-Rhap-(1→2)]-β-D-Glup; -26-*O*-β-D-Glup	254
	Compound 3, $C_{57}H_{94}O_{28}$, ESI-MS: 1179 [M−MeOH + H]$^+$, $[\alpha]_D^{21}$ −1.80° (*c* 1.67, MeOH)	(20*S*, 22*R*, 25*S*)-22-Methoxy-furost-5-ene-3,26-diol (**136**)	3-*O*-β-D-Xylp-(1→3)-β-D-Glup-(1→4)-[α-L-Rhap-(1→2)]-β-D-Glup; -26-*O*-β-D-Glup	
	Compound 4, WAP $C_{50}H_{80}O_{21}$, 263–265°C, MALDI-MS: 1017 [M + H]$^+$, $[\alpha]_D^{20}$ −1.77° (*c* 1.70, MeOH)	Diosgenin (**49**)	3-*O*-β-D-Xylp-(1→3)-β-D-Glup-(1→4)-[α-L-Rhap-(1→2)]-β-D-Glup	
	Compound 5, WAP $C_{50}H_{80}O_{21}$, 263–265°C, MALDI-MS: 1017 [M + H]$^+$, $[\alpha]_D^{20}$ −1.77° (*c* 1.70, MeOH)	Yamogenin (**50**)	3-*O*-β-D-Xylp-(1→3)-β-D-Glup-(1→4)-[α-L-Rhap-(1→2)]-β-D-Glup	
Calamus insignis (Palmae)	Compound 2, AS, $C_{57}H_{92}O_{26}$, FAB-MS: 1215 [M + Na]$^+$, $[\alpha]_D^{24}$ −73.1° (*c* 0.82, Pyr)	Diosgenin (**49**)	3-*O*-β-D-Glup-(1→4) α-L-Rhap-(1→4)-β-D-Glup-(1→4)-[α-L-Rhap-(1→2)]-β-D-Glup	129

	Compound 3, AS, $C_{51}H_{82}O_{21}$, FAB-MS: 1053 [M+Na]$^+$, $[\alpha]_D^{23}$, $-80.4°$ (c 1.45, Pyr)	Yamogenin (**50**)	3-O-α-L-Rhap-$(1 \rightarrow 4)$-β-D-Glup-$(1 \rightarrow 4)$-[α-L-Rhap-$(1 \rightarrow 2)$]-β-D-Glup	
	Compound 4, AS, $C_{57}H_{92}O_{26}$, FAB-MS: 1233 [M+Na]$^+$, $[\alpha]_D^{24}$ $-150.5°$ (c 1.57, Pyr)	(25R)-Furost-5-ene-3β,22α,26-triol (**127**)	3-O-α-L-Rhap-$(1 \rightarrow 4)$-β-D-Glup-$(1 \rightarrow 4)$-[α-L-Rhap-$(1 \rightarrow 2)$]-β-D-Glup	
	Compound 5, AS, $C_{51}H_{82}O_{21}$, FAB-MS: 1053 [M+Na]$^+$, $[\alpha]_D^{24}$ $-36.9°$ (c 1.0, Pyr)	22-Epiyamogenin (**51**)	3-O-α-L-Rhap-$(1 \rightarrow 4)$-β-D-Glup-$(1 \rightarrow 4)$-[α-L-Rhap-$(1 \rightarrow 2)$]-β-D-Glup	
Camassia leichtlinii (Liliaceae)	Compound 2, AS, $C_{57}H_{92}O_{28}$, FAB-MS: 1247 [M+Na]$^+$, $[\alpha]_D^{25}$ $-36.0°$ (c 0.1, MeOH)	Neohecogenin (**31**)	-3-O-{β-D-Glup-$(1 \rightarrow 2)$-O-[O-α-L-Rhap-$(1 \rightarrow 4)$-β-D-Glup-$(1 \rightarrow 3)$]-O-β-D-Glup-$(1 \rightarrow 4)$-β-D-Galp}	*255*
	Compound 3, AS, $C_{57}H_{94}O_{28}$, FAB-MS: 1249 [M+Na]$^+$, $[\alpha]_D^{25}$ $-36.0°$ (c 0.1, MeOH)	(25R)-5α-Spirostane-3β,15α-diol (**14**)	-3-O-{β-D-Glup-$(1 \rightarrow 2)$-O-[O-α-L-Rhap-$(1 \rightarrow 4)$-β-D-Glup-$(1 \rightarrow 3)$]-O-β-D-Glup-$(1 \rightarrow 4)$-β-D-Galp}	

Table 1 (continued)

Plant name and family	Glycoside, physical nature, mp (°C), Mol. formula, Mol. wt. (m/z), $[\alpha]_D$	Aglycone/sapogenin	Sugar with linkage	Reference
	Compound 5, AS, $C_{63}H_{104}O_{33}$, FAB-MS: 1411 $[M+Na]^+$, $[\alpha]_D^{25}$ −40.0° (c 0.1, MeOH)	(25R)-5α-Spirostane-3β,15α-diol (**14**)	3-O-{β-D-Glup-$(1 \rightarrow 3)$-O-β-D-Glup-$(1 \rightarrow 2)$-O-[O-α-L-Rhap-$(1 \rightarrow 4)$-β-D-Glup-$(1 \rightarrow 3)$]-O-β-D-Glup-$(1 \rightarrow 4)$-β-D-Galp}	
	Compound 6, AS, $C_{63}H_{104}O_{33}$, FAB-MS: 1411 $[M+Na]^+$, $[\alpha]_D^{25}$ −44.0° (c 0.1, MeOH)	Rockogenin (**13**)	3-O-{β-D-Glup-$(1 \rightarrow 3)$-O-β-D-Glup-$(1 \rightarrow 2)$-O-[O-α-L-Rhap-$(1 \rightarrow 4)$-β-D-Glup-$(1 \rightarrow 3)$]-O-β-D-Glup-$(1 \rightarrow 4)$-β-D-Galp}	
	Compound 7, AS, $C_{63}H_{102}O_{34}$, FAB-MS: 1425 $[M+Na]^+$, $[\alpha]_D^{25}$ −34.0° (c 0.1, MeOH)	(25R)-5α-Spirostane-3β,15α-diol-12-one (**15**)	3-O-{β-D-Glup-$(1 \rightarrow 3)$-O-β-D-Glup-$(1 \rightarrow 2)$-O-[O-α-L-Rhap-$(1 \rightarrow 4)$-β-D-Glup-$(1 \rightarrow 3)$]-O-β-D-Glup-$(1 \rightarrow 4)$-β-D-Galp}	

	Compound 8, AS, $C_{70}H_{118}O_{38}$, FAB-MS: 1565 $[M-H]^-$, $[\alpha]_D^{25}$ −44.0° (c 0.1, MeOH)	(25R)-22ξ-Methoxy-5α-furostane-3β,26-diol (**93**)	-3-O-{β-D-Glup-(1→3)-O-β-D-Glup-(1→2)-O-[O-α-L-Rhap-(1→4)-β-D-Glup-(1→3)]-O-β-D-Glup-(1→4)-β-D-Galp}
	Compound 11, AS, $C_{38}H_{62}O_{13}$, FAB-MS: 725 $[M-H]^-$, $[\alpha]_D^{25}$ −28.0° (c 0.1, MeOH)	Chlorogenin (**8**)	-6-O-β-D-Xylp-(1→2)-O-β-D-Glup
Cestrum nocturnum (Solanaceae)	Compound 1, AS, $C_{62}H_{100}O_{34}$, HR-TOF-MS: 1411.5925 $[M+Na]^+$, $[\alpha]_D^{28}$ −48.0° (c 0.1, MeOH)	(24S, 25S)-Spirost-5-ene-2α,3β,24-triol (**70**)	-3-O-β-D-Glup-(1→3)-O-β-D-Glup-(1→2)-O-[β-D-Xylp-(1→3)]-O-β-D-Glup-(1→4)-β-D-Galp; -24-O-β-D-Glup
	Compound 2, AP, $C_{63}H_{104}O_{34}$, HR-TOF-MS: 1427.6340 $[M+Na]^+$, $[\alpha]_D^{27}$ −60.0° (c 0.1, MeOH)	(25R)-22α-Methoxyfurost-5-ene-2α,3β,26-triol (**141**)	-3-O-β-D-Glup-(1→3)-O-β-D-Glup-(1→2)-O-[β-D-Xylp-(1→3)]-O-β-D-Glup-(1→4)-β-D-Galp; -26-O-β-D-Glup
	Compound 3, AP, $C_{62}H_{100}O_{33}$, HR-TOF-MS: 1395.6025 $[M+Na]^+$, $[\alpha]_D^{28}$ −46.0° (c 0.1, MeOH)	(25R)-Furosta-5,20(22)-diene-2α,3β,26-triol (**146**)	-3-O-β-D-Glup-(1→3)-O-β-D-Glup-(1→2)-O-[β-D-Xylp-(1→3)]-O-β-D-Glup-(1→4)-β-D-Galp; -26-O-β-D-Glup

183

Table 1 (continued)

Plant name and family	Glycoside, physical nature, mp (°C), Mol. formula, Mol. wt. (m/z), $[\alpha]_D$	Aglycone/sapogenin	Sugar with linkage	Reference
	Compound 4, AP, $C_{50}H_{80}O_{24}$, FAB-MS: 1087 $[M+Na]^+$, $[\alpha]_D^{24}$ −70.8° (c 0.13, CHCl$_3$-MeOH, 1:1)	(25R)-Spirost-5-ene-2α, 3β,17α-triol (**69**)	-3-*O*-β-D-Glup- (1→2)-*O*-[β-D-Xylp- (1→3)]-*O*-β-D-Glup- (1→4)-β-D-Galp	*180*
	Compound 6, AP, $C_{56}H_{90}O_{29}$, FAB-MS: 1249 $[M+Na]^+$, $[\alpha]_D^{24}$ −60.0° (c 0.13, CHCl$_3$-MeOH, 1:1)	(25R)-Spirost-5-ene-2α, 3β,15β-triol (**68**)	-3-*O*-β-D-Glup- (1→3)-*O*-β-D-Glup- (1→2)-*O*-[β-D-Xylp- (1→3)]-*O*-β-D-Glup- (1→4)-β-D-Galp	
	Compound 7, AP, $C_{56}H_{90}O_{29}$, FAB-MS: 1249 $[M+Na]^+$, $[\alpha]_D^{24}$ −57.0° (c 0.2, CHCl$_3$-MeOH, 1:1)	(25R)-Spirost-5-ene-2α, 3β,17α-triol (**69**)	-3-*O*-β-D-Glup- (1→3)-*O*-β-D-Glup- (1→2)-*O*-[β-D-Xylp- (1→3)]-*O*-β-D-Glup- (1→4)-β-D-Galp	
	Compound 9, AP, $C_{51}H_{82}O_{21}$, FAB-MS: 1053 $[M+Na]^+$, $[\alpha]_D^{24}$ −93.3° (c 0.12, CHCl$_3$-MeOH, 1:1)	Yuccagenin (**54**)	-3-*O*-α-L-Rhap- (1→2)-*O*-[*O*-α-L-Rhap-(1→4)-α-L-Rhap-(1→4)]-β-D-Glup	
C. sendtenerianum	Compound 1, AS, $C_{39}H_{60}O_{14}$, HR-FAB-MS: 753.4081 $[M+H]^+$, $[\alpha]_D^{25}$ −70.7° (c 0.2, MeOH)	Spirosta-5,25(27)-diene-1β,2α,3β-triol (**86**)	-3-*O*-α-L-Rhap- (1→2)-β-D-Galp	*130*
	Compound 2, AS, $C_{39}H_{62}O_{14}$, HR-FAB-MS: 777.4000 $[M+Na]^+$, $[\alpha]_D^{25}$ −57.1° (c 0.14, MeOH)	(25R)-Spirost-5-ene-1β, 2α,3β-triol (**65**)	-3-*O*-α-L-Rhap- (1→2)-β-D-Galp	

	Compound 3, AS, $C_{39}H_{62}O_{14}$, HR-FAB-MS: 777.4076 [M+Na]$^+$, $[\alpha]_D^{25}$ −54.2° (c 0.43, MeOH)	5α-Spirost-25(27)-ene-1β,2α,3β-triol (**48**)	-3-O-α-L-Rhap-(1→2)-β-D-Galp
	Compound 4, AS, $C_{39}H_{64}O_{14}$, HR-FAB-MS: 779.4198 [M+Na]$^+$, $[\alpha]_D^{25}$ −56.4° (c 0.11, MeOH)	(25R)-5α-Spirostane-1β, 2α,3β-triol (**18**)	-3-O-α-L-Rhap-(1→2)-β-D-Galp
	Compound 5, AS, $C_{45}H_{70}O_{19}$, HR-FAB-MS: 937.4405 [M+Na]$^+$, $[\alpha]_D^{25}$ −124.4° (c 0.25, MeOH)	Spirosta-5,25(27)-diene-1β,2α,3β-triol (**86**)	-3-O-α-L-Rhap-(1→2)-O-[β-D-Glup-(1→4)]-β-D-Galp
	Compound 2, AS, $C_{33}H_{50}O_{11}$, HR-FAB-MS: 645.3245 [M+Na]$^+$, $[\alpha]_D^{25}$ −47.6° (c 0.25, MeOH)	Spirosta-5,25(27)-diene-1β,2α,3β,12β-tetrol (**89**)	-3-O-β-D-Galp	256
Cordyline stricta (Agavaceae)	Compound 1, AS, $C_{44}H_{70}O_{16}$, FAB-MS: 853 [M−H]$^−$, $[\alpha]_D^{27}$ −54.5° (c 0.29, MeOH)	1β-Hydroxy-crabbogenin (**48**)	-1-O-{O-α-L-Rhap-(1→2)-O-[β-D-Xylp-(1→3)]-β-D-Fucp}	257
	Compound 2, AS, $C_{43}H_{70}O_{16}$, FAB-MS: 881 [M+K]$^+$, $[\alpha]_D^{28}$ −46.5° (c 0.16, MeOH)	(25S)-5α-Spirostane-1β, 3α-diol (**169**)	-1-O-{O-α-L-Rhap-(1→2)-O-[β-D-Xylp-(1→3)]-β-D-Xylp}
	Compound 3, AS, $C_{43}H_{68}O_{16}$, FAB-MS: 879 [M+K]$^+$, $[\alpha]_D^{27}$ −72.6° (c 0.52, MeOH)	1β-Hydroxy-crabbogenin (**48**)	-1-O-{O-α-L-Rhap-(1→2)-O-[β-D-Xylp-(1→3)]-β-D-Xylp}
	Compound 4, AS, $C_{51}H_{84}O_{22}$, FAB-MS: 1047 [M−H]$^−$, $[\alpha]_D^{20}$ −20.0° (c 0.11, MeOH)	22ξ-Methoxy-5α-furost-25(27)-ene-1β,3β,26-triol (**118**)	-1-O-{O-α-L-Rhap-(1→2)-O-[β-D-Xylp-(1→3)]-β-D-Fucp}; -26-O-β-D-Glup

Plant name and family	Glycoside, physical nature, mp (°C), Mol. formula, Mol. wt. (m/z), $[\alpha]_D$	Aglycone/sapogenin	Sugar with linkage	Reference
	Compound 5, AS, $C_{51}H_{82}O_{22}$, FAB-MS: 1045 [M−H]⁻, $[\alpha]_D^{27}$ −24.0° (c 0.12, MeOH)	22ξ-Methoxy-furosta-5,25(27)-diene-1β,3,3,β,26-triol (**123**)	-1-*O*-{*O*-α-L-Rhap-(1→2)-*O*-[β-D-Xylp-(1→3)]-β-D-Fucp}; -26-*O*-β-D-Glup	
Costus spicattus (Costaceae)	Compound 1, colorless needles, $C_{51}H_{84}O_{22}$, 222–224°C, LSI-MS: 1047 [M−H]⁻, $[\alpha]_D^{20}$ −102.0° (c 0.001, MeOH)	(25R)-22α-Methoxy-furost-5-ene-3β,26-diol (**131**)	-3-*O*-β-D-Apiof-(1→2)-*O*-[6-deoxy-α-L-Manp-(1→4)]-β-D-Glup	258
Dioscorea cayenensis (Dioscoreaceae)	Compound 1, WAP, $C_{57}H_{92}O_{27}$, HR-ESI-MS: 1231.5697 [M+Na]⁺, $[\alpha]_D^{20}$ +80.0° (c 0.025, MeOH)	(25R)-20,22-*seco*-Furost-5-ene-3β,26-diol-20,22-dione (**149**)	3-*O*-α-L-Rhap-(1→4)-α-L-Rhap-(1→4)-[α-L-Rhap-(1→2)]-β-D-Glup-26-*O*-β-D-Glup	216
D. cayenensis	Compound 1	25(R)-22ξ-Methoxy-furost-5-ene-3β,26-diol (**133**)	3-*O*-α-L-Rhap-(1→4)-α-L-Rhap-(1→4)-[α-L-Rhap-(1→2)]-β-D-Glup; -26-*O*-β-D-Glup	215
D. panthaica	Dioscoreside A, WAS, $C_{51}H_{82}O_{24}$, 178–180° (dec), ESI-MS: 1077 [M−H]⁻, $[\alpha]_D^{25}$ −50.2° (c 0.003, Pyr)	(25R)-20,22-*seco*-Furost-5-ene-3β,26-diol-20,22-dione (**149**)	-3-*O*-β-D-Glup-(1→3)-*O*-α-L-Rhap-(1→2)-β-D-Glup; -26-*O*-β-D-Glup	259

	Dioscoreside B, AS, $C_{51}H_{82}O_{24}$, 186–188° (dec), ESI-MS: 1077 [M−H]−, $[\alpha]_D^{25}$ −69.1° (c 0.005, Pyr)	(23S, 25R)-20,22-seco-Furost-5-ene-3β,23,26-triol-20,22-dione (**150**)	-3-O-β-D-Glup-(1→4)-α-L-Rhap-(1→2)-α-L-Rhap; -26-O-β-D-Glup	260
D. panthaica	Dioscoreside C	(23S, 25R)-23-Methoxy-furosta-5,20(22)-diene-3β,26-diol (**144**)	-3-O-α-L-Rhap-(1→2)-[α-L-Rhap-(1→4)]-β-D-Glup-26-O-β-D-Glup	
	Dioscoreside D	(25R)-20,22-seco-Furost-5-ene-3β,26-diol 20,22-dione (**149**)	-3-O-α-L-Rhap-(1→2)-[α-L-Rhap-(1→4)]-β-D-Glup; -26-O-β-D-Glup	
D. polygonoides	Compound 1, AS, $C_{39}H_{62}O_{14}$, HR-ESI-MS: 755.4213 [M+H]+, $[\alpha]_D^{26}$ −114.0° (c 0.1, MeOH)	(23S, 24R, 25S)-Spirost-5-ene-3β,23,24-triol (**75**)	-3-O-α-L-Rhap-(1→2)-β-D-Glup	261
	Compound 2, AS, $C_{39}H_{62}O_{15}$, HR-ESI-MS: 771.4191 [M+H]+, $[\alpha]_D^{27}$ −98.0° (c 0.1, MeOH)	(23S, 25R)-Spirost-5-ene-3β,12α,17α,23-tetrol (**81**)	-3-O-α-L-Rhap-(1→2)-β-D-Glup	
	Compound 3, AS, $C_{39}H_{62}O_{15}$, HR-ESI-MS: 793.3992 [M+Na]+, $[\alpha]_D^{26}$ −84.0° (c 0.1, MeOH)	(23S, 25R)-Spirost-5-ene-3β,14α,17α,23-tetrol (**82**)	-3-O-α-L-Rhap-(1→2)-β-D-Glup	
D. pseudojaponica	Compound 1, AS, $C_{58}H_{96}O_{26}$, 189–190°C (dec), ESI-MS: 1231 [M+Na]+, $[\alpha]_D^{16}$ −86.4° (c 0.05, MeOH)	(25R)-22α-Methoxy-furost-5-ene-3β,26-diol (**131**)	-3-O-α-L-Rhap-(1→2)-O-{[α-L-Rhap-(1→4)]-O-[α-L-Rhap-(1→4)]}-β-D-Glup; -26-O-β-D-Glup	262

Table 1 (continued)

Plant name and family	Glycoside, physical nature, mp (°C), Mol. formula, Mol. wt. (m/z), $[\alpha]_D$	Aglycone/sapogenin	Sugar with linkage	Reference
	Compound 4, AS, $C_{51}H_{82}O_{22}$, 243–245°C, ESI-MS: 1037 $[M+Na]^+$, $[\alpha]_D^{16}$ –104.7° (c 0.05, MeOH)	Diosgenin (**49**)	-3-O-α-L-Rhap-$(1\rightarrow2)$-O-{[α-L-Rhap-$(1\rightarrow4)$]-O-[α-L-Rhap-$(1\rightarrow4)$]}-β-D-Glup	
Disporopsis pernyi (Liliaceae)	Disporoside A, WAP, $C_{45}H_{74}O_{18}$, FAB-MS: 902 $[M]^-$, $[\alpha]_D^{23}$ –0.5° (c 0.40, Pyr)	Smilagenin (**35**)	-3-O-β-D-Glup-$(1\rightarrow2)$-[β-D-Glup-$(1\rightarrow6)$]-β-D-Glup	263
	Disporoside B, WAP, $C_{55}H_{94}O_{14}$, FAB-MS: 977 $[M-H]^-$, $[\alpha]_D^{23}$ –54.2° (c 0.40, Pyr)	Smilagenin (**35**)	-3-O-β-D-Glup-$(1\rightarrow2)$-[6-O-hexadecanoyl-β-D-Glup-$(1\rightarrow6)$]-β-D-Glup	
	Disporoside C, WAP, $C_{45}H_{76}O_{19}$, FAB-MS: 919 $[M-H]^-$, $[\alpha]_D^{23}$ –40.9° (c 0.20, Pyr)	(22R, 25R)-5β-Furostane-3β,22,26-triol (**113**)	-3-O-β-D-Glup-$(1\rightarrow2)$-β-D-Glup; -26-O-β-D-Glup	
	Disporoside D, WAP, $C_{51}H_{86}O_{24}$, FAB-MS: 1082 $[M]^-$, $[\alpha]_D^{23}$ –43.7° (c 0.40, Pyr)	(22R, 25R)-5β-Furostane-3β,22,26-triol (**113**)	-3-O-β-D-Glup-$(1\rightarrow2)$-[β-D-Glup-$(1\rightarrow6)$]-β-D-Glup; -26-O-β-D-Glup	
Dracaena angustifolia (Dracaenaceae)	Namonin A, AS, $C_{57}H_{84}O_{26}$, FAB-MS: 1183 $[M-H]^-$, $[\alpha]_D^{25}$ –65.7° (c 0.5, MeOH)	(23S, 24R)-Spirosta-5,25(27)-diene-1β,3,3β,23,24-tetrol (**91**)	-1-O-{[2,3,4-tri-O-acetyl-α-L-Rhap-$(1\rightarrow2)$]-[3-O-acetyl-β-D-Xylp-$(1\rightarrow3)$]-α-L-Arap}; -24-O-β-D-Fucp	196

	Namonin B, AS, $C_{57}H_{84}O_{26}$, FAB-MS: 1183 $[M-H]^-$, $[\alpha]_D^{25}$ −68.8° (c 0.8, MeOH)	(23S, 24R)-Spirosta-5, 25(27)-diene-1β,3β,23, 24-tetrol (**91**)	-1-O-[2,3,4-tri-O-acetyl-α-L-Rhap-(1→2)]-[4-O-acetyl-β-D-Xylp-(1→3)]-α-L-Arap}; -24-O-β-D-Fucp
	Namonin C, AS, $C_{44}H_{68}O_{18}$, FAB-MS: 883 $[M-H]^-$, $[\alpha]_D^{25}$ −109.7° (c 0.9, MeOH)	(23S, 24R)-Spirosta-5, 25(27)-diene-1β,3β,23, 24-tetrol (**91**)	-1-O-[α-L-Rhap-(1→2)]-α-L-Arap}; -24-O-β-D-Fucp
	Namonin D, AS, $C_{46}H_{70}O_{19}$, FAB-MS: 925 $[M-H]^-$, $[\alpha]_D^{25}$ −58.7° (c 0.3, MeOH)	(23S, 24R)-Spirosta-5, 25(27)-diene-1β,3β,23, 24-tetrol (**91**)	-1-O-{[4-O-acetyl-α-L-Rhap-(1→2)]-α-L-Arap}; -24-O-β-D-Fucp
	Namonin E, AS, $C_{51}H_{80}O_{22}$, FAB-MS: 1067 $[M+Na]^+$, $[\alpha]_D^{25}$ −49.4° (c 0.5, MeOH)	(25R)-Furosta-5,20(22)-diene-1β,3β,26-triol (**145**)	-1-O-{α-L-Rhap-(1→2)-[β-D-Xylp-(1→3)]-4-O-acetyl-α-L-Arap}; -26-O-β-D-Glup
	Namonin F, AS, $C_{44}H_{68}O_{19}$, FAB-MS: 899 $[M-H]^-$, $[\alpha]_D^{25}$ −18.4° (c 0.1, MeOH)	20,22-seco-Furosta-5,25(27)-diene-1β,3β,26-triol-20,22-dione (**151**)	-1-O-{α-L-Rhap-(1→2)-β-D-Arap}; -26-O-β-D-Glup
D. cochinchinensis (Agavaceae)	Dracaenoside I, AS, $C_{45}H_{70}O_{17}$, FAB-MS: 881 $[M-H]^-$, $[\alpha]_D^{20}$ −62.50° (c 0.2, MeOH)	Sceptrumgenin (**84**)	3-O-{O-α-L-Rhap-(1→2)-O-[β-D-Glup-(1→3)]-β-D-Glup}	*264*

Table 1 (continued)

Plant name and family	Glycoside, physical nature, mp (°C), Mol. formula, Mol. wt. (m/z), $[\alpha]_D$	Aglycone/sapogenin	Sugar with linkage	Reference
	Dracaenoside J, AS, $C_{45}H_{72}O_{19}$, FAB-MS: 915 [M−H]$^-$, $[\alpha]_D^{20}$ −85.0° (c 0.2, MeOH)	Spirost-5-ene-3β,14,27-triol (**73**)	-3-O-{O-α-L-Rhap-(1→2)-O-[β-D-Glup-(1→4)]-α-L-Rhap}	
	Dracaenoside K, WAP, $C_{45}H_{72}O_{18}$, FAB-MS: 900 [M]$^+$, $[\alpha]_D^{20}$ −100.0° (c 0.2, MeOH)	Spirosta-5-ene-3β,14,24-triol (**72**)	-3-O-{O-α-L-Rhap-(1→2)-O-[β-D-Glup-(1→4)-α-L-Rhap}	
	Dracaenoside L, WAP, $C_{45}H_{72}O_{19}$, FAB-MS: 915 [M−H]$^-$, $[\alpha]_D^{28}$ −66.67° (c 0.2, MeOH)	Spirost-5-ene-3β,14,24-triol (**72**)	-3-O-{O-α-L-Rhap-(1→2)-O-[β-D-Glup-(1→3)]-β-D-Glup}	
	Dracaenoside R, AS, $C_{45}H_{72}O_{19}$, FAB-MS: 915 [M−H]$^-$, $[\alpha]_D^{20}$ −75.13° (c 0.2, MeOH)	(22S,25S)-22,25-Epoxy-furost-5-ene-3β,14α,26,27-tetrol (**153**)	-3-O-{O-α-L-Rhap-(1→2)-O-[β-D-Glup-(1→4)]-α-L-Rhap}	
D. concinna	Compound 10, AS, $C_{46}H_{76}O_{18}$, FAB-MS: 915 [M−H]$^-$, $[\alpha]_D^{27}$ −45.0° (c 0.12, MeOH)	22ξ-Methoxy-5α-furost-25(27)-ene-1β,3α,26-triol (**117**)	-1-O-{O-α-L-Rhap-(1→2)-O-β-D-Fucp}; -26-O-β-D-Glup	265
	Compound 11, AS, $C_{45}H_{74}O_{18}$, FAB-MS: 901 [M−H]$^-$, $[\alpha]_D^{27}$ −37.5° (c 0.41, MeOH)	22ξ-Methoxy-5α-furost-25(27)-ene-1β,3α,26-triol (**117**)	-1-O-{O-α-L-Rhap-(1→2)-O-α-L-Arap}; -26-O-β-D-Glup	
	Compound 12, AS, $C_{46}H_{76}O_{19}$, FAB-MS: 931 [M−H]$^-$, $[\alpha]_D^{29}$ −64.0° (c 0.1, MeOH)	22ξ-Methoxy-5α-furost-25(27)-ene-1β,3α,4α,26-tetrol (**119**)	-1-O-{O-α-L-Rhap-(1→2)-O-β-D-Fucp}; -26-O-β-D-Glup	

D. draco	Compound 13, AS, $C_{46}H_{76}O_{19}$, FAB-MS: 931 $[M-H]^-$, $[\alpha]_D^{29}$ $-64.8°$ (c 0.11, MeOH)	22ξ-Methoxy-5α-furost-25(27)-ene-1β,3β,4α,26-tetrol (**120**)	-1-*O*-{*O*-α-L-Rhap-(1→2)-*O*-β-D-Fucp}; -26-*O*-β-D-Glup	266
	Draconin A, AS, $C_{44}H_{64}O_{17}$, FAB-MS: 865 $[M+H]^+$, $[\alpha]_D^{20}$ $-70.0°$ (c 1.5, EtOH)	(23*S*, 24*S*)-Spirosta-5,25(27)-diene-1β,3β,23,24-tetrol (**91**)	-1-*O*-(2,3,4-tri-*O*-acetyl-α-L-Rhap)-(1→2)-α-L-Arap}	
	Draconin B, AS, $C_{42}H_{62}O_{16}$, HR-FAB-MS: 846.3954 $[M+Na+H]^+$, $[\alpha]_D^{20}$ $-100.0°$ (c 2.6, EtOH)	(23*S*, 24*S*)-Spirosta-5,25(27)-diene-1β,3β,23,24-tetrol (**91**)	-1-*O*-{*O*-(2,3-di-*O*-acetyl-α-L-Rhap)-(1→2)-α-L-Arap}	
	Draconin C, AS, $C_{40}H_{60}O_{15}$, HR-FAB-MS: 780.3865 $[M+Na]^+$, $[\alpha]_D^{20}$ $-85.0°$ (c 1.5, EtOH)	(23*S*, 24*S*)-Spirosta-5,25(27)-diene-1β,3β,23,24-tetrol (**91**)	-1-*O*-{*O*-(2-*O*-acetyl-α-L-Rhap)-(1→2)-α-L-Arap}	
	Compound 5, AS, $C_{50}H_{74}O_{21}$, FAB-MS: 1009 $[M-H]^-$, $[\alpha]_D^{24}$ $-62.0°$ (c 0.1, MeOH)	(23*S*, 24*S*)-Spirosta-5,25(27)-diene-1β,3β,23,24-tetrol (**91**)	-1-*O*-{*O*-(2,3,4-tri-*O*-acetyl-α-L-Rhap)-(1→2)-α-L-Arap}; -24-*O*-β-D-Fucp	174
	Compound 6, AS, $C_{38}H_{58}O_{14}$, FAB-MS: 737 $[M-H]^-$, $[\alpha]_D^{26}$ $-88.0°$ (c 0.1, MeOH)	(23*S*, 24*S*)-Spirosta-5,25(27)-diene-1β,3β,23,24-tetrol (**91**)	-1-*O*-{*O*-α-L-Rhap-(1→2)-*O*-α-L-Arap}	
	Compound 7, AS, $C_{40}H_{60}O_{15}$, FAB-MS: 779 $[M-H]^-$, $[\alpha]_D^{26}$ $-78.0°$ (c 0.1, MeOH)	(23*S*, 24*S*)-Spirosta-5,25(27)-diene-1β,3β,23,24-tetrol (**91**)	-1-*O*-{*O*-(4-*O*-acetyl-α-L-Rhap)-(1→2)-α-L-Arap}	
	Compound 8, AS, $C_{38}H_{58}O_{13}$, FAB-MS: 721 $[M-H]^-$, $[\alpha]_D^{24}$ $-90.0°$ (c 0.1, MeOH)	(23*S*)-Spirosta-5,25(27)-diene-1β,3β,23-triol (**87**)	-1-*O*-{*O*-α-L-Rhap-(1→2)-*O*-α-L-Arap}	

Table 1 (continued)

Plant name and family	Glycoside, physical nature, mp (°C), Mol. formula, Mol. wt. (m/z), $[\alpha]_D$	Aglycone/sapogenin	Sugar with linkage	Reference
	Compound 9, AS, $C_{40}H_{60}O_{14}$, FAB-MS: 763 [M−H]$^-$, $[\alpha]_D^{26}$ −62.0° (c 0.1, MeOH)	(23S)-Spirosta-5,25(27)-diene-1β,3β,23-triol (**87**)	-1-O-{O-(4-O-acetyl-α-L-Rhap)-(1 → 2)-α-L-Arap}	
	Icogenin, AS, $C_{46}H_{76}O_{18}$, FAB-MS: 907 [M+Na-OMe]$^+$, $[\alpha]_D^{20}$ −61.2° (c 0.04, EtOH)	(25S)-22ζ-Methoxy-furost-5-ene-3β,26-diol (**134**)	-3-O-α-L-Rhap-(1 → 2)-[β-D-Glup-(1 → 3)]-β-D-Glup	192
D. surculosa	Surculoside A, AS, $C_{44}H_{70}O_{18}$, HR-FAB-MS: 909.4465 [M+Na]$^+$, $[\alpha]_D^{25}$ −106.0° (c 0.1, CHCl$_3$–MeOH, 1:1)	(24S, 25R)-Spirost-5-ene-1β,3β,24-triol (**67**)	-1-O-β-D-Fucp; -3-O-β-D-Apiof-(1 → 4)-β-D-Glup	267
	Surculoside B, AS, $C_{39}H_{62}O_{14}$, HR-FAB-MS: 777.4054 [M+Na]$^+$, $[\alpha]_D^{25}$ −114.0° (c 0.1, CHCl$_3$–MeOH, 1:1)	(24S, 25R)-Spirost-5-ene-1β,3β,24-triol (**67**)	-1-O-β-D-Fucp; -24-O-β-D-Glup	
	Surculoside C, AS, $C_{45}H_{72}O_{18}$, HR-FAB-MS: 923.4642 [M+Na]$^+$, $[\alpha]_D^{25}$ −136.0° (c 0.1, CHCl$_3$–MeOH, 1:1)	(24S, 25R)-Spirost-5-ene-1β,3β,24-triol (**67**)	-1-O-α-L-Rhap-(1 → 2)-O-β-D-Fucp; -24-O-β-D-Glup	
	Surculoside D, AS, $C_{40}H_{66}O_{15}$, FAB-MS: 785 [M−H]$^-$, $[\alpha]_D^{25}$ −104° (c 0.1, CHCl$_3$–MeOH, 1:1)	(25S)-22α-Methoxy-furost-5-ene-1β,3β,26-triol (**140**)	-1-O-β-D-Glup; -26-O-β-D-Glup	

	Compound 1, AS, $C_{39}H_{62}O_{14}$, FAB-MS: 777 $[M+Na]^+$, $[\alpha]_D^{26}$ −90.0° (c 0.1, MeOH)	(24S, 25R)-3α,5α-Cyclospirostane-1β,6β,24-triol (**159**)	-1-O-β-D-Fucp; -24-O-β-D-Glup	*54*
	Compound 2, AS, $C_{39}H_{62}O_{15}$, FAB-MS: 793 $[M+Na]^+$, $[\alpha]_D^{26}$ −42.0° (c 0.1, MeOH)	(24S, 25R)-3α,5α-Cyclospirostane-1β,6β,24-triol (**159**)	-1-O-β-D-Glup; -24-O-β-D-Glup	
	Compound 3, AS, $C_{40}H_{66}O_{15}$, FAB-MS: 785 $[M−H]^−$, $[\alpha]_D^{26}$ −42.0° (c 0.1, MeOH)	(25S)-22α-Methoxy-3α,5α-cyclofurostane-1β,6β,26-triol (**160**)	-1-O-β-D-Glup; -26-O-β-D-Glup	
	Compound 4, AS, $C_{40}H_{66}O_{14}$, FAB-MS: 769 $[M−H]^−$, $[\alpha]_D^{26}$ −56.0° (c 0.1, MeOH)	(25S)-22α-Methoxy-3α,5α-cyclofurostane-1β,6β,26-triol (**160**)	-1-O-β-D-Fucp; -26-O-β-D-Glup	
Furcraea selloa var. *marginata* (Agavaceae)	Furcrea furostatin, AS, $C_{69}H_{116}O_{38}$, ESI-MS: 1553 $[M+H]^+$, $[\alpha]_D^{20}$ +96.6° (c 0.1, H_2O)	(25R)-5α-Furostane-3β,22ξ,26-triol (**100**)	-3-O-[α-L-Rhap-(1→4)-β-D-Glup-(1→3)-{β-D-Glup-(1→3)-β-D-Glup-(1→2)}-β-D-Glup-(1→4)-β-D-Galp]; -26-O-β-D-Glup	*268*
	Furcrea furostatin methyl ether, AS, $C_{70}H_{118}O_{38}$, ESI-MS: 1567 $[M+H]^+$, $[\alpha]_D^{20}$ −218.4° (c 0.10, H_2O)	(25R)-22-Methoxy-5α-furostane-3β,22ξ,26-triol (**93**)	-3-O-[α-L-Rhap-(1→4)-β-D-Glup-(1→3)-{β-D-Glup-(1→3)-β-D-Glup-(1→2)}-β-D-Glup-(1→4)-β-D-Galp]; -26-O-β-D-Glup	

Table 1 (continued)

Plant name and family	Glycoside, physical nature, mp (°C), Mol. formula, Mol. wt. (*m/z*), $[\alpha]_D$	Aglycone/sapogenin	Sugar with linkage	Reference
Fructus Trichosanthis (Cucurbitaceae) and *Bulbus Allii Macrostemi* (Alliaceae)	Compound 1, AP, $C_{39}H_{62}O_{14}$, 265–266°C, FAB-MS: 755 $[M+H]^+$	Schidegeragenin C (**46**)	-3-*O*-β-D-Glup-$(1 \rightarrow 2)$-β-D-Galp	269
	Compound 2, AP, $C_{39}H_{64}O_{16}$, 175–176°C, FAB-MS: 787 $[M-H]^-$	5β-Furost-25(27)-ene-1β,3β,6β,22α,26-pentol (**121**)	-3-*O*-β-D-Galp; -26-*O*-β-D-Glup	
Helleborus orientalis (Ranunculaceae)	Compound 4, AS, $C_{50}H_{76}O_{22}$, FAB-MS: 1027 $[M-H]^-$, $[\alpha]_D^{26}$ −64.0° (*c* 0.1, MeOH)	(23*S*)-Spirosta-5,25(27)-diene-1β,3β,23-triol (**87**)	-1-*O*-β-D-Apiof-$(1 \rightarrow 3)$-*O*-(4-*O*-acetyl-α-L-Rhap)-$(1 \rightarrow 2)$-*O*-[β-D-Xylp-$(1 \rightarrow 3)$]-α-L-Arap	270
	Compound 5, AS, $C_{50}H_{76}O_{23}$, FAB-MS: 1067 $[M+Na]^+$, $[\alpha]_D^{25}$ −104.0° (*c* 0.1, MeOH)	(23*S*, 24*S*)-Spirosta-5,25(27)-diene-1β,3β,23,24-tetrol (**91**)	-1-*O*-β-D-Apiof-$(1 \rightarrow 3)$-*O*-(4-*O*-acetyl-α-L-Rhap)-$(1 \rightarrow 2)$-*O*-[β-D-Xylp-$(1 \rightarrow 3)$]-α-L-Arap	
	Compound 6, AS, $C_{52}H_{78}O_{25}$, FAB-MS: 1125 $[M+Na]^+$, $[\alpha]_D^{28}$ −78.0° (*c* 0.1, MeOH)	(23*S*, 24*S*)-21-Acetoxy-spirosta-5,25(27)-diene-1β,3β,23,24-tetrol (**92**)	-1-*O*-β-D-Apiof-$(1 \rightarrow 3)$-*O*-(4-*O*-acetyl-α-L-Rhap)-$(1 \rightarrow 2)$-*O*-[β-D-Xylp-$(1 \rightarrow 3)$]-α-L-Arap	

Source	Compound data	Name	Substituents	Ref
	Compound 7, AS, $C_{58}H_{88}O_{30}$, FAB-MS: 1287 $[M+Na]^+$, $[\alpha]_D^{28}$ −76.0° (c 0.1, MeOH)	(23S, 24S)-21-Acetoxy-spirosta-5,25(27)-diene-1β,3β,23,24-tetrol (**92**)	-1-O-β-D-Apiof-$(1\to 3)$-O-(4-O-acetyl-α-L-Rhap)-$(1\to 2)$-O-[β-D-Xylp-$(1\to 3)$]-α-L-Arap; -24-O-β-D-Glup	
	Compound 8, AS, $C_{58}H_{88}O_{29}$, FAB-MS: 1271 $[M+Na]^+$, $[\alpha]_D^{26}$ −72.0° (c 0.1, MeOH)	(23S, 24S)-21-Acetoxy-spirosta-5,25(27)-diene-1β,3β,23,24-tetrol (**92**)	-1-O-β-D-Apiof-$(1\to 3)$-O-(4-O-acetyl-α-L-Rhap)-$(1\to 2)$-O-[β-D-Xylp-$(1\to 3)$]-α-L-Arap; -24-O-β-D-Quinp	
H. viridis L. (Ranunculaceae)	Compound 1, AP, $C_{54}H_{88}O_{24}$, ESI-MS: 1106 $[M-CH_3+H]^+$, $[\alpha]_D^{25}$ −46.0° (c 0.05, MeOH)	(25R)-22α-Methoxy-furost-5-ene-3β,26-diol (**131**)	-3-O-β-D-Glup-$(1\to 3)$-O-[6-O-acetyl-β-D-Glup-$(1\to 3)$]-O-β-D-Glup; -26-O-α-L-Rhap	271
	Compound 2, AP, $C_{52}H_{86}O_{23}$, ESI-MS: 1064 $[M-CH_3+H]^+$, $[\alpha]_D^{25}$ −70.0° (c 0.1, MeOH)	(25R)-22α-Methoxy-furost-5-ene-3β,26-diol (**131**)	-3-O-β-D-Glup-$(1\to 3)$-O-β-D-Glup-$(1\to 3)$-O-β-D-Glup; -26-O-α-L-Rhap	
Hemerocallis furva var. *kwanso* (Liliaceae)	Hemeroside A, WP, $C_{38}H_{62}O_{15}$, 120–125°C, HR-FAB-MS: 781.4001 $[M+Na]^+$, $[\alpha]_D^{26}$ −17.6° (c 0.9, MeOH)	(24S)-Hydroxy-neotokorogenin (**40**)	-1-O-α-L-Arap; -24-O-β-D-Glup	272
	Hemeroside B, colorless needles, $C_{50}H_{82}O_{23}$, 287–290°C, HR-FAB-MS: 1073.5144 $[M+Na]^+$, $[\alpha]_D^{26}$ −56.0° (c 1.5, Pyr)	Isorhodeasapogenin (**36**)	-3-O-β-D-Glup-$(1\to 3)$-[β-D-Xylp-$(1\to 2)$]-β-D-Glup-$(1\to 4)$-β-D-Galp	

Table 1 (continued)

Plant name and family	Glycoside, physical nature, mp (°C), Mol. formula, Mol. wt. (m/z), $[\alpha]_D$	Aglycone/sapogenin	Sugar with linkage	Reference
Hosta sieboldii (Liliaceae)	Compound 13, AS, $C_{45}H_{72}O_{20}$, FAB-MS: 931 $[M-H]^-$, $[\alpha]_D^{25}$ −50.0° (c 0.1, CHCl$_3$–MeOH, 1:1)	Manogenin (**6**)	-3-O-{O-β-D-Glup-$(1\rightarrow 2)$-O-β-D-Glup-$(1\rightarrow 4)$-β-D-Galp}	169
	Compound 14, AS, $C_{45}H_{70}O_{20}$, FAB-MS: 929 $[M-H]^-$, $[\alpha]_D^{25}$ −60.0° (c 0.1, CHCl$_3$–MeOH, 1:1)	9,11-Dehydro-manogenin (**7**)	-3-O-{O-β-D-Glup-$(1\rightarrow 2)$-O-β-D-Glup-$(1\rightarrow 4)$-β-D-Galp}	
	Compound 15, AS, $C_{56}H_{88}O_{28}$, FAB-MS: 1207 $[M-H]^-$, $[\alpha]_D^{25}$ −26.0° (c 0.1, MeOH)	9,11-Dehydro-manogenin (**7**)	-3-O-{O-β-D-Glup-$(1\rightarrow 2)$-O-[O-α-L-Rhap-$(1\rightarrow 4)$-β-D-Xylp-$(1\rightarrow 3)$]-O-β-D-Glup-$(1\rightarrow 4)$-β-D-Galp}	
	Compound 16, AS, $C_{57}H_{94}O_{30}$, FAB-MS: 1257 $[M-H]^-$, $[\alpha]_D^{25}$ −60.0° (c 0.1, MeOH)	(25R)-22α-Methoxy-5α-furostane-2α,3β,26-triol-12-one (**96**)	-3-O-{O-β-D-Glup-$(1\rightarrow 2)$-O-[β-D-Xylp-$(1\rightarrow 3)$]-O-β-D-Glup-$(1\rightarrow 4)$-β-D-Galp}-26-O-β-D-Glup	
	Compound 17, AS, $C_{57}H_{92}O_{30}$, FAB-MS: 1255 $[M-H]^-$, $[\alpha]_D^{25}$ −24.0° (c 0.1, MeOH)	(25R)-22α-Methoxy-5α-furost-9-ene-2α,3β,26-triol-12-one (**97**)	-3-O-{O-β-D-Glup-$(1\rightarrow 2)$-O-[β-D-Xylp-$(1\rightarrow 3)$]-O-β-D-Glup-$(1\rightarrow 4)$-β-D-Galp}-26-O-β-D-Glup	

	Compound 18, AS, $C_{39}H_{64}O_{14}$, FAB-MS: 755 [M−H]⁻, $[\alpha]_D^{25}$ −84.0° (c 0.1, CHCl₃-MeOH, 1:1)	(25R)-5α-Spirostane-2α,3β,12β-triol (**21**)	-3-O-{O-α-L-Rhap-(1→2)-β-D-Galp}
Lilium candidum (Liliaceae)	Compound 1, AS, $C_{45}H_{72}O_{17}$, FAB-MS: 883 [M−H]⁻, $[\alpha]_D^{29}$ −89.6° (c 0.27, MeOH)	Diosgenin (**49**)	-3-O-α-L-Rhap-(1→2)-O-[β-D-Glup-(1→6)]-β-D-Glup	273
	Compound 2, AS, $C_{45}H_{72}O_{18}$, FAB-MS: 899 [M−H]⁻, $[\alpha]_D^{27}$ −44.2° (c 0.12, MeOH-H₂O, 1:1)	Isonarthogenin (**63**)	-3-O-α-L-Rhap-(1→2)-O-[β-D-Glup-(1→6)]-β-D-Glup
	Compound 3, AS, $C_{45}H_{72}O_{18}$, FAB-MS: 899 [M−H]⁻, $[\alpha]_D^{26}$ −41.5° (c 0.28, Pyr)	(23S, 25R)-Spirost-5-ene-3β,23-diol (**59**)	-3-O-α-L-Rhap-(1→2)-O-[β-D-Glup-(1→6)]-β-D-Glup
	Compound 4, AS, $C_{46}H_{74}O_{18}$, FAB-MS: 913 [M−H]⁻, $[\alpha]_D^{27}$ −47.1° (c 0.14, MeOH-H₂O, 1:1)	(25R, 26R)-26-Methoxyspirost-5-ene-3β-diol (**83**)	-3-O-α-L-Rhap-(1→2)-O-[β-D-Glup-(1→6)]-β-D-Glup
	Compound 5, AS, $C_{46}H_{74}O_{19}$, FAB-MS: 929 [M−H]⁻, $[\alpha]_D^{27}$ −42.1° (c 0.14, MeOH-H₂O, 1:1)	(25R, 26R)-26-Methoxyspirost-5-ene-17α,3β-diol (**58**)	-3-O-α-L-Rhap-(1→2)-O-[β-D-Glup-(1→6)]-β-D-Glup
	Compound 6, AS, $C_{52}H_{86}O_{23}$, FAB-MS: 1077 [M−H]⁻, $[\alpha]_D^{29}$ −69.0° (c 0.29, MeOH)	(25R)-22ξ-Methoxy-furost-5-ene-3β,26-diol (**133**)	-3-O-α-L-Rhap-(1→2)-O-[β-D-Glup-(1→6)]-β-D-Glup
	Compound 2, AS, $C_{46}H_{74}O_{19}$, FAB-MS: 929 [M−H]⁻, $[\alpha]_D^{27}$ −42.1° (c 0.14, MeOH:H₂O, 1:1)	(25R, 26R)-26-Methoxyspirost-5-ene-17α,3β-diol (**58**)	-3-O-{O-α-L-Rhap-(1→2)-O-[β-D-Glup-(1→4)]-β-D-Glup}	274

Table 1 (continued)

Plant name and family	Glycoside, physical nature, mp (°C), Mol. formula, Mol. wt. (m/z), $[\alpha]_D$	Aglycone/sapogenin	Sugar with linkage	Reference
	Compound 3, AS, $C_{48}H_{76}O_{20}$, FAB-MS: 971 [M−H]$^-$, $[\alpha]_D^{27}$ −36.7° (c 0.15, MeOH:H$_2$O, 1:1)	(25R, 26R)-26-Methoxy-spirost-5-ene-17α,3β-diol (**58**)	3-O-{O-α-L-Rhap-(1→2)-O-[6-O-acetyl-β-D-Glup-(1→4)]-β-D-Glup}	
	Compound 7, AS, $C_{40}H_{64}O_{14}$, FAB-MS: 967 [M−H]$^-$, $[\alpha]_D^{26}$ −30.6° (c 0.26, Pyr)	(25R, 26R)-26-Methoxy-spirost-5-ene-17α,3β-diol (**58**)	3-O-{O-α-L-Rhap-(1→2)-O-β-D-Glup}	
	Compound 8, AS, $C_{51}H_{82}O_{23}$, FAB-MS: 1061 [M−H]$^-$, $[\alpha]_D^{26}$ −10.6° (c 0.25, Pyr)	Isonarthogenin (**63**)	3-O-{O-β-D-Glup-(1→3)-O-α-L-Rhap-(1→2)-O-[β-D-Glup-(1→4)]-β-D-Glup}	
Ophiopogon japonicus (Liliaceae)	Ophiojaponin C, colorless needles, 215–217°C, $C_{46}H_{72}O_{18}$, FAB-MS: 885 [M−H]$^-$, $[\alpha]_D^{12.3}$ −77.3° (c 0.51, MeOH)	Ophiopogenin (**71**)	3-O-[α-L-Arap-(1→2)]-β-D-Xylp-(1→4)-β-D-Glup	275
Ornithogalum thyrsoides (Liliaceae)	Compound 1, AS, $C_{33}H_{54}O_9$, HR-ESI-MS: 595.3836 [M+H]$^+$, $[\alpha]_D^{25}$ −68.0° (c 0.1, MeOH)	(25R)-5α-Spirostane-1β,3β-diol (**3**)	1-O-β-D-Glup	276
	Compound 3, AS, $C_{43}H_{68}O_{16}$, HR-ESI-MS: 863.4404 [M+Na]$^+$, $[\alpha]_D^{25}$ −54.0° (c 0.1, MeOH)	Ruscogenin (**52**)	1-O-α-L-Arap-(1→2)-O-[β-D-Xylp-(1→3)]-α-L-Arap	
	Compound 4, AS, $C_{38}H_{60}O_{13}$, HR-ESI-MS: 747.3935 [M+Na]$^+$, $[\alpha]_D^{25}$ −72.0° (c 0.1, MeOH)	(24S, 25S)-Spirost-5-ene-1β,3β,24-triol (**66**)	1-O-α-L-Rhap-(1→2)-α-L-Arap	

	Compound 5, AS, $C_{43}H_{68}O_{17}$, HR-ESI-MS: 879.4302 [M+Na]$^+$, $[\alpha]_D^{26}$ −48.0° (c 0.1, MeOH)	(24S, 25S)-Spirost-5-ene-1β,3β,24-triol (**66**)	-1-O-α-L-Arap-(1→2)-O-[β-D-Xylp-(1→3)]-α-L-Arap
	Ornithosaponin A, AS, $C_{38}H_{58}O_{15}$, HR-ESI-MS: 777.3704 [M+Na]$^+$, $[\alpha]_D^{26}$ −90.0° (c 0.1, MeOH)	(23S, 24S, 25S)-1β,3β,23,24-Tetrahydroxy-spirost-5-en-15-one (**80**)	-1-O-α-L-Rhap-(1→2)-α-L-Arap
	Ornithosaponin B, AS, $C_{44}H_{68}O_{19}$, HR-ESI-MS: 923.4244 [M+Na]$^+$, $[\alpha]_D^{26}$ −114.0° (c 0.1, MeOH)	(23S, 24S, 25S)-1β,3β,23,24-Tetrahydroxy-spirost-5-en-15-one (**80**)	-1-O-α-L-Rhap-(1→2)-α-L-Arap; -24-(6-deoxy-β-D-Gulp)
	Ornithosaponin C, AS, $C_{49}H_{76}O_{23}$, HR-ESI-MS: 1055.4716 [M+Na]$^+$, $[\alpha]_D^{26}$ −70.0° (c 0.1, MeOH)	(23S, 24S, 25S)-1β,3β,23,24-Tetrahydroxy-spirost-5-en-15-one (**80**)	-1-O-α-L-Rhap-(1→2)-O-[β-D-Xylp-(1→3)]-α-L-Arap; -24-(6-deoxy-β-D-Gulp)
	Ornithosaponin D, AS, $C_{55}H_{82}O_{26}$, HR-ESI-MS: 1181.5088 [M+Na]$^+$, $[\alpha]_D^{26}$ −96.0° (c 0.1, MeOH)	(23S, 24S, 25S)-1β,3β,23,24-Tetrahydroxy-spirost-5-en-15-one (**80**)	-1-{O-(2,3,4-tri-O-acetyl-α-L-Rhap)-(1→2)-O-[β-D-Xylp-(1→3)]-α-L-Arap}; -24-(6-deoxy-β-D-Gulp)
			277
Polianthes tuberosa (Agavaceae)	Polianthoside B, WAP, $C_{56}H_{91}O_{27}$, HR-FAB-MS: 1195.5709 [M−H]$^−$, $[\alpha]_D^{18.3}$ −52.04° (c 0.022, Pyr)	Tigogenin (**1**)	-3-O-β-D-Xylp-(1→3)-β-D-Glup-(1→2)-[β-D-Glup-(1→3)]-β-D-Glup-(1→4)-β-D-Galp
			135

Table 1 (continued)

Plant name and family	Glycoside, physical nature, mp (°C), Mol. formula, Mol. wt. (m/z), $[\alpha]_D$	Aglycone/sapogenin	Sugar with linkage	Reference
	Polianthoside C, WAP, $C_{57}H_{94}O_{28}$, HR-FAB-MS: 1226.5870 [M−H]$^-$, $[\alpha]_D^{19.8}$ −32.79° (c 0.018, Pyr)	Tigogenin (**1**)	-3-O-β-D-Glup- $(1 \rightarrow 3)$-β-D-Glup- $(1 \rightarrow 2)$-[β-D-Glup- $(1 \rightarrow 3)$]-β-D-Glup- $(1 \rightarrow 4)$-β-D-Galp	
	Polianthoside D, WAP, $C_{56}H_{92}O_{29}$, HR-FAB-MS: 1227.5735 [M−H]$^-$, $[\alpha]_D^{18.1}$ −23.2° (c 0.047, Pyr)	(25R)-5α-Furostane-3β, 22α,26-triol-12-one (**101**)	-3-O-β-D-Glup- $(1 \rightarrow 2)$-[β-D-Xylp- $(1 \rightarrow 3)$]-β-D-Glup- $(1 \rightarrow 4)$-β-D-Galp; -26-O-β-D-Glup	
	Polianthoside E, WAP, $C_{61}H_{100}O_{33}$, HR-FAB-MS: 1359.6039 [M−H]$^-$, $[\alpha]_D^{18.1}$ −23.53° (c 0.034, Pyr)	(25R)-5α-Furostane-3β, 22α,26-triol-12-one (**101**)	-3-O-β-D-Xylp- $(1 \rightarrow 3)$-β-D-Glup- $(1 \rightarrow 2)$-[β-D-Xylp- $(1 \rightarrow 3)$]-β-D-Glup- $(1 \rightarrow 4)$-β-D-Galp; -26-O-β-D-Glup	
	Polianthoside F, WAP, $C_{61}H_{102}O_{32}$, HR-FABMS: 1345.6194 [M−H]$^-$, $[\alpha]_D^{19.8}$ −37.18° (c 0.039, Pyr)	(25R)-5α-Furostane-3β, 22α,26-triol-12-one (**101**)	-3-O-β-D-Xylp- $(1 \rightarrow 3)$-β-D-Glup- $(1 \rightarrow 2)$-[β-D-Xylp- $(1 \rightarrow 3)$]-β-D-Glup- $(1 \rightarrow 4)$-β-D-Galp; -26-O-β-D-Glup	

	Polianthoside G, WAP, $C_{62}H_{104}O_{33}$, HR-FAB-MS: 1375.6420 [M−H]⁻, $[\alpha]_D^{19.7}$ −35.26° (c 0.039, Pyr)	(25R)-5α-Furostane-3β, 22α,26-triol-12-one (**101**)	3-*O*-β-D-Xylp-(1→3)-β-D-Glup-(1→2)-[β-D-Glup-(1→3)]-β-D-Glup-(1→4)-β-D-Galp; 26-*O*-β-D-Glup	*173*
	Compound 2, AS, $C_{55}H_{90}O_{27}$, FAB-MS: 1181 [M−H]⁻, $[\alpha]_D^{27}$ −42.0° (c 0.1, MeOH)	Chlorogenin (**8**)	3-*O*-β-D-Xylp-(1→3)-*O*-β-D-Glup-(1→2)-*O*-[β-D-Xylp-(1→3)]-*O*-β-D-Glup-(1→4)-β-D-Galp	
	Compound 3, AS, $C_{55}H_{88}O_{27}$, FAB-MS: 1179 [M−H]⁻, $[\alpha]_D^{26}$ −20.0° (c 0.1, MeOH)	Hecogenin (**32**)	3-*O*-β-D-Xylp-(1→3)-*O*-β-D-Glup-(1→2)-*O*-[β-D-Xylp-(1→3)]-*O*-β-D-Glup-(1→4)-β-D-Galp	
	Compound 4, AS, $C_{55}H_{86}O_{27}$, FAB-MS: 1177 [M−H]⁻, $[\alpha]_D^{27}$ −35.5° (c 0.22, MeOH)	(25R)-5α-Spirost-9-ene-3β-ol-12-one (**33**)	3-*O*-β-D-Xylp-(1→3)-*O*-β-D-Glup-(1→2)-*O*-[β-D-Xylp-(1→3)]-*O*-β-D-Glup-(1→4)-β-D-Galp	
Polygonatum kingianum (Convallariaceae)	(25S)-Kingianoside D, WAP, $C_{45}H_{72}O_{19}$, HR-FAB-MS: 939.4609 [M+Na]⁺, $[\alpha]_D^{20}$ −18.5° (c 0.065, Pyr)	(25S)-Furost-5-ene-3β, 22ξ,26-triol-12-one (**129**)	3-*O*-β-D-Glup-(1→4)-β-D-Fucp; 26-*O*-β-D-Glup	*40*
	(25S)-Kingianoside C, WAP, $C_{45}H_{72}O_{20}$, HR-FAB-MS: 955.4542 [M+Na]⁺, $[\alpha]_D^{20}$ −42.3° (c 0.265, Pyr)	(25S)-Furost-5-ene-3β, 22ξ,26-triol-12-one (**129**)	3-*O*-β-D-Glup-(1→4)-β-D-Galp-26-*O*-β-D-Glup	
	(25R,22ξ)-Hydroxywattinoside C WAP, $C_{45}H_{74}O_{20}$, HR-FAB-MS: 957.4686 [M+Na]⁺, $[\alpha]_D^{20}$ −33° (c 0.41, Pyr)	(25R)-Furost-5-ene-1β,3β,22ξ,26-tetrol (**137**)	3-*O*-β-D-Glup-(1→4)-β-D-Galp; 26-*O*-β-D-Glup	

Table 1 (continued)

Plant name and family	Glycoside, physical nature, mp (°C), Mol. formula, Mol. wt. (m/z), $[\alpha]_D$	Aglycone/sapogenin	Sugar with linkage	Reference
	Kingianoside E, WAP, $C_{51}H_{82}O_{25}$, HR-FAB-MS: 1117.5034 $[M+Na]^+$, $[\alpha]_D^{20}$ −28.9° (c 0.405, Pyr)	(25R)-Furost-5-ene-3β, 22ξ,26-triol-12-one (**130**)	3-O-β-D-Glup-(1→2)-β-D-Glup-(1→4)-β-D-Galp; -26-O-β-D-Glup	
	(25S)-Kingianoside E, WAP, $C_{51}H_{82}O_{25}$, HR-FAB-MS: 1117.5056 $[M+Na]^+$, $[\alpha]_D^{20}$ −30.5° (c 0.4, Pyr)	(25S)-Furost-5-ene-3β, 22ξ,26-triol-12-one (**129**)	3-O-β-D-Glup-(1→2)-β-D-Glup-(1→4)-β-D-Galp; -26-O-β-D-Glup	
	Kingianoside F, WAP, $C_{51}H_{84}O_{25}$, HR-FAB-MS: 1119.5201 $[M+Na]^+$, $[\alpha]_D^{20}$ −36.7° (c 0.302, Pyr)	(25R)-Furost-5-ene-1β, 3β,22ξ,26-tetrol (**137**)	3-O-β-D-Glup-(1→2)-β-D-Glup-(1→4)-β-D-Galp; -26-O-β-D-Glup	
	(25S)-Kingianoside F, WAP, $C_{51}H_{84}O_{25}$, HR-FAB-MS: 1119.5219 $[M+Na]^+$, $[\alpha]_D^{20}$ −43.0° (c 0.33, Pyr)	(25S)-Furost-5-ene-1β, 3β,22ξ,26-tetrol (**138**)	3-O-β-D-Glup-(1→2)-β-D-Glup-(1→4)-β-D-Galp; -26-O-β-D-Glup	
P. sibiricum	Neosibiricoside A, AP, $C_{47}H_{74}O_{21}$, ESI-MS: 997 $[M+Na]^+$, $[\alpha]_D^{20}$ −31.5° (c 0.24, Pyr)	(23S, 24R, 25R)-1β-Acetoxy-spirost-5-ene-3β,23,24-triol (**76**)	-3-O-β-D-Glup-(1→2)-β-D-Glup-(1→4)-β-D-Fucp	184
	Neosibiricoside B, WAP, ESI-MS: 1113 $[M+Na]^+$, $[\alpha]_D^{20}$ −36.3° (c 0.14, Pyr-MeOH)	Ruscogenin 1-acetate (**53**)	-3-O-β-D-Glup-(1→2)-[β-D-Xylp-(1→3)]-β-D-Glup-(1→4)-β-D-Galp	

	Neosibiricoside C, WAP, ESI-MS: 1097 [M+Na]⁺, [α]$_D^{20}$ −76.4° (c 0.09, Pyr-MeOH)	Yamogenin (**50**)	-3-O-β-D-Glup-(1→2)-[β-D-Xylp-(1→3)]-β-D-Glup-(1→4)-2-O-acetyl-β-D-Galp	
P. zanlanscianense (Liliaceae)	Polygonatoside A, AP, C₄₅H₇₀O₁₉, HR-FAB-MS: 913.4379 [M−H]⁻, [α]$_D^{20}$ −24.51° (c 0.148, Pyr)	(25S)-Spirost-5-ene-3β, 27-diol-12-one (**64**)	-3-O-β-D-Glup-(1→4)-β-D-Fucp; -27-O-β-D-Glup	278
	Polygonatoside B, AP, C₄₅H₇₀O₂₀, HR-FAB-MS: 929.4425 [M−H]⁻, [α]$_D^{20}$ −19.19° (c 0.052, Pyr)	(25S)-Spirost-5-ene-3β, 27-diol-12-one (**64**)	-3-O-β-D-Glup-(1→4)-β-D-Galp; -27-O-β-D-Glup	
	Polygonatoside C, AP, C₃₉H₅₉O₁₅, HR-FAB-MS: 767.3870 [M−H]⁻, [α]$_D^{20}$ −48.43° (c 0.035, Pyr)	(23S, 25S)-Spirost-5-ene-3β,23,27-triol-12-one (**78**)	-3-O-β-D-Glup-(1→4)-β-D-Fucp	
	Polygonatoside D, AP, C₄₅H₇₂O₁₈, HR-FAB-MS: 899.4781 [M−H]⁻, [α]$_D^{20}$ −50.31° (c 0.014, Pyr)	Isonarthogenin (**63**)	-3-O-[α-L-Rhap-(1→4)]-β-D-Glup; -27-O-β-D-Glup	
Ruscus aculeatus (Liliaceae)	Compound 6, AS, C₃₈H₅₇O₁₅NaS, FAB-MS: 1061 [M−H]⁻, [α]$_D^{26}$ −84.0° (c 0.1, MeOH)	Neoruscogenin (**85**)	-1-O-{O-α-L-Rhap-(1→2)-4-O-sulpho-α-L-Arap}	65
	Compound 7, AS, FAB-MS: 947 [M-Na-OMe-H]⁻, [α]$_D^{26}$ −52.0° (c 0.1, MeOH)	22ξ-Methoxy-furosta-5,25(27)-diene-1β,3β,26-triol (**123**)	-1-O-{O-α-L-Rhap-(1→2)-4-O-sulpho-α-L-Arap}; -26-O-β-D-Glup	
	Compound 8, AS, FAB-MS: 1021 [M−Na−H]⁻, [α]$_D^{26}$ −34.0° (c 0.1, MeOH)	22ξ-Methoxy-furosta-5,25(27)-diene-1β,3β,26-triol (**123**)	-1-O-{O-α-L-Rhap-(1→2)-3-O-acetyl-4-O-sulpho-α-L-Arap}; -26-O-β-D-Glup	

Table 1 (*continued*)

Plant name and family	Glycoside, physical nature, mp (°C), Mol. formula, Mol. wt. (*m/z*), $[\alpha]_D$	Aglycone/sapogenin	Sugar with linkage	Reference
	Compound 9, AS, $C_{53}H_{84}O_{21}$, FAB-MS: 1055 [M−H]$^-$, $[\alpha]_D^{26}$ −44.0° (c 0.1, MeOH)	22ξ-Methoxy-furosta-5,25(27)-diene-1β,3β,26-triol (**123**)	-1-*O*-{*O*-α-L-Rhap-(1→2)-3-*O*-acetyl-4-*O*-[(2S, 3S)-2-hydroxy-3-methyl pentanoyl]-α-L-Arap}; -26-*O*-β-D-Glup	
	Compound 10, AS, $C_{46}H_{70}O_{15}$, FAB-MS: 861 [M−H]$^-$, $[\alpha]_D^{26}$ −40.0° (c 0.1, MeOH)	Neoruscogenin (**85**)	-1-*O*-{*O*-α-L-Rhap-(1→2)-3-*O*-acetyl-4-*O*-[(2S, 3S)-2-hydroxy-3-methyl pentanoyl]-α-L-Arap}	
	Compound 11, AS, FAB-MS: 941 [M−H]$^-$, $[\alpha]_D^{26}$ −38.0° (c 0.1, MeOH)	22ξ-Methoxy-furosta-5,25(27)-diene-1β,3β,26-triol (**123**)	-1-*O*-{*O*-α-L-Rhap-(1→2)-4-*O*-acetyl-α-L-Arap}; -26-*O*-β-D-Glup	
	Compound 12, AS, FAB-MS: 909 [M−H]$^-$, $[\alpha]_D^{26}$ −54.0° (c 0.1, MeOH)	Neoruscogenin (**85**)	-1-*O*-{*O*-β-D-Glup-(1→3)-*O*-α-L-Rhap-(1→2)-4-*O*-acetyl-α-L-Arap}	
	Compound 1, AS, $C_{50}H_{78}O_{23}$, FAB-MS: 1045 [M−H]$^-$, $[\alpha]_D^{26}$ −50.0° (c 0.1, MeOH)	(23S)-Spirosta-5,25(27)-diene-1β,3β,23-triol (**87**)	-1-*O*-{*O*-β-D-Glup-(1→3)-*O*-α-L-Rhap-(1→2)-α-L-Arap}; -23-*O*-β-D-Glup	279

Compound 2, AS, $C_{44}H_{68}O_{18}$; FAB-MS: 883 [M−H]⁻, $[\alpha]_D^{26}$ −52.0° (c 0.10, MeOH)	(23S)-Spirosta-5,25(27)-dien-1β,3β,23-triol (**87**)	-1-O-{O-α-L-Rhap-(1→2)-α-L-Arap}; -23-O-β-D-Glup
Compound 1, AS, $C_{39}H_{62}O_{13}$; FAB-MS: 737 [M−H]⁻, $[\alpha]_D^{25}$ −74.0° (c 0.1, MeOH)	Ruscogenin (**52**)	-1-O-{O-α-L-Rhap-(1→2)-β-D-Galp}
Compound 2, AS, $C_{41}H_{64}O_{14}$; FAB-MS: 779 [M−H]⁻, $[\alpha]_D^{25}$ −76.0° (c 0.1, MeOH)	Ruscogenin (**52**)	-1-O-{O-α-L-Rhap-(1→2)-6-O-acetyl-β-D-Galp}
Compound 3, AS, $C_{43}H_{66}O_{15}$; FAB-MS: 821 [M−H]⁻, $[\alpha]_D^{25}$ −62.0° (c 0.1, MeOH)	Ruscogenin (**52**)	-1-O-{O-α-L-Rhap-(1→2)-4,6-di-O-acetyl-β-D-Galp}
Compound 4, AS, $C_{45}H_{68}O_{16}$; FAB-MS: 863 [M−H]⁻, $[\alpha]_D^{25}$ −78.0° (c 0.1, MeOH)	Ruscogenin (**52**)	-1-O-{O-α-L-Rhap-(1→2)-3,4,6-tri-O-acetyl-β-D-Galp}
Compound 5, AS, $C_{45}H_{72}O_{18}$; FAB-MS: 899 [M−H]⁻, $[\alpha]_D^{25}$ −46.0° (c 0.1, MeOH)	Ruscogenin (**52**)	-1-O-{O-β-D-Glup-(1→3)-O-α-L-Rhap-(1→2)-β-D-Galp}
Compound 6, AS, $C_{49}H_{76}O_{20}$; FAB-MS: 983 [M−H]⁻, $[\alpha]_D^{25}$ −50.0° (c 0.1, MeOH)	Ruscogenin (**52**)	-1-O-{O-β-D-Glup-(1→3)-O-α-L-Rhap-(1→2)-4,6-di-O-acetyl-β-D-Galp}
Compound 7, AS, $C_{46}H_{76}O_{19}$; FAB-MS: 931 [M−H]⁻, $[\alpha]_D^{25}$ −36.0° (c 0.1, MeOH)	(25R)-22ξ-Methoxy-furost-5-ene-1β,3β,26-triol (**139**)	-1-O-{O-α-L-Rhap-(1→2)-β-D-Galp}

63

Table 1 (continued)

Plant name and family	Glycoside, physical nature, mp (°C), Mol. formula, Mol. wt. (m/z), $[\alpha]_D$	Aglycone/sapogenin	Sugar with linkage	Reference
	Compound 8, AS, $C_{48}H_{78}O_{20}$, FAB-MS: 973 [M−H]$^-$, $[\alpha]_D^{25}$ −46.0° (c 0.1, MeOH)	(25R)-22ξ-Methoxy-furost-5-ene-1β,3β,26-triol (**139**)	1-O-{O-α-L-Rhap-(1→2)-6-O-acetyl-β-D-Galp}	
	Compound 9, AS, $C_{52}H_{82}O_{22}$, FAB-MS: 1057 [M−H]$^-$, $[\alpha]_D^{25}$ −160.0° (c 0.1, MeOH)	(25R)-22ξ-Methoxy-furost-5-ene-1β,3β,26-triol (**139**)	1-O-{O-α-L-Rhap-(1→2)-3,4,6-tri-O-acetyl-β-D-Galp}	
	Compound 10, AS, $C_{52}H_{86}O_{24}$, FAB-MS: 1093 [M−H]$^-$, $[\alpha]_D^{25}$ −98.0° (c 0.1, MeOH)	(25R)-22ξ-Methoxy-furost-5-ene-1β,3β,26-triol (**139**)	1-O-{O-β-D-Glup-(1→3)-O-α-L-Rhap-(1→2)-β-D-Galp}	
	Compound 11, AS, $C_{58}H_{92}O_{27}$, FAB-MS: 1219 [M−H]$^-$, $[\alpha]_D^{25}$ −40.0° (c 0.1, MeOH)	(25R)-22ξ-Methoxy-furost-5-ene-1β,3β,26-triol (**139**)	1-O-{O-β-D-Glup-(1→3)-O-α-L-Rhap-(1→2)-3,4,6-tri-O-acetyl-β-D-Galp}; -26-O-β-D-Glup	
	Compound 1, AS, $C_{39}H_{62}O_{14}$, FAB-MS: 753 [M−H]$^-$, $[\alpha]_D^{25}$ −44.0° (c 0.1, MeOH)	(23S, 25R)-Spirost-5-ene-3β,23-diol (**59**)	23-O-{O-β-D-Glup-(1→6)-β-D-Glup}	280
Sansevieria ehrenbergii (Agavaceae)	Sansevierin A, AS, $C_{39}H_{62}O_{13}$, 226–230°C, HR-FAB-MS: 745.4319 [M+Li]$^+$, $[\alpha]_D^{24}$ −104.0° (c 0.62, MeOH)	(25R)-Spirost-5-ene-3β,7α-diol (**55**)	3-O-[α-L-Rhap-(1→2)]-β-D-Glup	37
	Sansevistatin 1, AP, $C_{45}H_{70}O_{16}$, 267–269°C, HR-APCI-MS: 867.4761 [M+H]$^+$, $[\alpha]_D^{24}$ −102.0° (c 0.4, MeOH)	Sceptrumgenin (**84**)	3-O-[α-L-Rhap-(1→2)]-[α-L-Rhap-(1→4)]-β-D-Glup	

	Sansevistatin 2, AP, $C_{44}H_{70}O_{16}$, 280–282°C, HR-FAB-MS: 861.4864 $[M+Li]^+$, $[\alpha]_D^{24}$ −87.1° (c 0.68, Pyr)	Diosgenin (**49**)	3-O-{α-L-Arap-(1→4)-[α-L-Rhap-(1→2)]-β-D-Glup}
Smilacina atropurpurea (Convallariaceae)	Atropuroside A, WAP, $C_{38}H_{60}O_{13}$, FAB-MS: 723 $[M+Li]^+$, $[\alpha]_D$ −76.5° (c 0.22, MeOH)	(25R)-Spirost-5-ene-1β,2α,3β-triol (**65**)	-1-O-α-L-Rhap-(1→2)-β-D-Xylp
	Atropuroside B, WAP, $C_{38}H_{60}O_{14}$, FAB-MS: 739 $[M-H]^-$, $[\alpha]_D$ −70.3° (c 0.38, MeOH)	(25R)-Spirost-5-ene-1β,2α,3β,17α-tetrol (**79**)	-1-O-α-L-Rhap-(1→2)-β-D-Xylp
	Atropuroside C, WAP, $C_{32}H_{50}O_{10}$, FAB-MS: 593 $[M-H]^-$, $[\alpha]_D$ −62.1° (c 0.33, MeOH)	(25R)-Spirost-5-ene-1β,2α,3β,17α-tetrol (**79**)	-1-O-β-D-Xylp
	Atropuroside D, WAP, $C_{33}H_{50}O_{10}$, FAB-MS: 605 $[M-H]^-$, $[\alpha]_D$ −55.3° (c 0.18, MeOH)	Spirosta-5,25(27)-diene-1β,2α,3β-triol (**86**)	-1-O-β-D-Galp
	Atropuroside E, WAP, $C_{33}H_{50}O_{11}$, FAB-MS: 621 $[M-H]^-$, $[\alpha]_D$ −42.3° (c 0.06, MeOH)	Spirosta-5,25(27)-diene-1β,2α,3β,23α-tetrol (**90**)	-1-O-β-D-Galp
	Atropuroside F, WAP, $C_{39}H_{62}O_{16}$, FAB-MS: 785 $[M-H]^-$, $[\alpha]_D$ −27.4° (c 0.57, MeOH)	Furosta-5,25(27)-diene-1β,2α,3β,22ξ,26-pentol (**126**)	-1-O-β-D-Galp; -26-O-β-D-Glup
	Atropuroside G, WAP, $C_{39}H_{62}O_{15}$, FAB-MS: 769 $[M-H]^-$, $[\alpha]_D$ −22.0° (c 0.28, MeOH)	Furosta-5,25(27)-diene-22ξ-methoxy-1β,2α,3β,26-tetrol (**125**)	-1-O-β-D-Xylp; -26-O-β-D-Glup

Table 1 (continued)

Plant name and family	Glycoside, physical nature, mp (°C), Mol. formula, Mol. wt. (m/z), $[\alpha]_D$	Aglycone/sapogenin	Sugar with linkage	Reference
Smilax medica (Smilacaceae)	Compound 1, WAP, $C_{51}H_{84}O_{23}$, FAB-MS: 1063 [M−H]$^-$, $[\alpha]_D^{20}$ −22.2° (c 0.135, MeOH)	Smilagenin (**35**)	-3-*O*-β-D-Glup-$(1 \rightarrow 6)$-[β-D-Glup-$(1 \rightarrow 2)$]-[β-D-Glup-$(1 \rightarrow 4)$]-β-D-Glup	38
	Compound 2, WAP, $C_{45}H_{74}O_{18}$, FAB-MS: 901 [M−H]$^-$, $[\alpha]_D^{20}$ −109.8° (c 0.085, MeOH)	Smilagenin (**35**)	-3-*O*-β-D-Glup-$(1 \rightarrow 6)$-[β-D-Glup-$(1 \rightarrow 4)$]-β-D-Glup	
	Compound 3, WAP, $C_{58}H_{98}O_{29}$, FAB-MS: 1257 [M−H]$^-$, $[\alpha]_D^{20}$ −34.3° (c 0.333, MeOH)	(25*S*)-22α-Methoxy-5β-furostane-3β,26-diol (**111**)	-3-*O*-β-D-Glup-$-(1 \rightarrow 6)$-[β-D-Glup-$(1 \rightarrow 2)$]-[β-D-Glup-$(1 \rightarrow 4)$]-β-D-Glup; -26-*O*-β-D-Glup	
Solanum abutiloides (Solanaceae)	Abutiloside L, WP, $C_{49}H_{80}O_{20}$, FAB-MS: 1061 [M−H]$^-$, $[\alpha]_D^{25}$ −107.1° (c 1.15, MeOH)	(22*S*, 25*S*)-22,25-Epoxy-furost-5-ene-3β,7β,26-triol (**155**)	-3-*O*-β-Chacotrioside; -26-*O*-β-D-Glup	39
	Abutiloside M, WP, FAB-MS: 1075 [M−H]$^-$, $[\alpha]_D^{25}$ −110.9° (c 0.37, MeOH)	(22*S*, 25*S*)-22,25-Epoxy-7β-methoxy-furost-5-ene-3β,26-diol (**156**)	-3-*O*-β-chacotrioside; -26-*O*-β-D-Glup	
	Abutiloside N, WP, FAB-MS: 1077 [M−H]$^-$, $[\alpha]_D^{25}$ −84.8° (c 0.24, MeOH)	(22*S*, 25*S*)-22,25-Epoxy-furost-5-ene-3β,7β,26-triol (**155**)	-3-*O*-β-solatrioside; -26-*O*-β-D-Glup	
S. anguivi	Anguivioside III, WP, $C_{44}H_{70}O_{18}$, FAB-MS: 910 [M+Na+H]$^+$, $[\alpha]_D^{26}$ −67.4° (c 0.6, MeOH)	(22*R*, 23*S*, 25*R*, 26*R*)-Spirost-5-ene-3β,23,26-triol (**77**)	-3-*O*-[β-D-Xylp-$(1 \rightarrow 3)$]-α-L-Rhap-$(1 \rightarrow 2)$-β-D-Glup	281

	Anguiviosides XI, WP, $C_{50}H_{80}O_{23}$, FAB-MS: 1072 $[M+Na+H]^+$, $[\alpha]_D^{26}$ −61.4° (c 0.6, MeOH)	(22R, 23S, 25R, 26S)-Furost-5-en-23,26-epoxide-3β,22α,26-triol (**158**)	-3-O-[β-D-Xylp-(1 → 3)]-α-L-Rhap-(1 → 2)-β-D-Glup	282
	Anguivioside A	(25R, 26R)-Spirost-5-ene-3β,26-diol (**62**)	-3-O-β-chacotrioside	
	Anguivioside B	(25R, 26R)-Spirost-5-ene-3β,26-diol (**62**)	-3-O-[4-O-maloyl-α-L-Rhap-(1 → 2)]-α-L-Rhap-(1 → 4)-β-D-Glup	
	Anguivioside C	(25R, 26R)-Spirost-5-ene-3β,26-diol (**62**)	-3-O-α-L-Rhap-(1 → 2)-[β-D-Xylp-(1 → 3)]-β-D-Glup	
S. chrysotrichum (Solanaceae)	Saponin SC-2, WAP, 239–241°C, $[\alpha]_D^{26}$ −49.0° (c 1.08, MeOH)	Chlorogenin (**8**)	-6-O-β-D-Xylp-(1 → 3)-β-D-Quinp	214
	Saponin SC-3, WAP, 167–168°C	Chlorogenin (**8**)	-6-O-β-D-Xylp	
	Saponin SC-4, WAP, 194–196°C	Chlorogenin (**8**)	-6-O-β-D-Quinp	
	Saponin SC-5	Chlorogenin (**8**)	-6-O-α-L-Rhap-(1 → 3)-β-D-Quinp	
	Saponin SC-6, WAP, 198–199°C	Chrysogenin (**23**)	-6-O-α-L-Rhap-(1 → 3)-β-D-Quinp	
S. hispidum	Saponin 1, WAP, $C_{33}H_{54}O_8$, 190–194°C, HR-FAB-MS: 579.7982 $[M+H]^+$, $[\alpha]_D^{25}$ −18.0° (c 0.002, Pyr)	Neochlorogenin (**9**)	-6-O-β-D-Quinp	213

Table 1 (continued)

Plant name and family	Glycoside, physical nature, mp (°C), Mol. formula, Mol. wt. (m/z), $[\alpha]_D$	Aglycone/sapogenin	Sugar with linkage	Reference
S. khasianum	Solakhasoside 1, AP, $C_{44}H_{69}O_{18}$, 250–252°C, FAB-MS: 885 $[M-H]^-$, $[\alpha]_D^{28}$ −50.0° (c 0.1, MeOH)	(23S, 25S)-Spirost-5-ene-3β,17α,23-triol (**74**)	-3-O-{α-L-Rhap-$(1\rightarrow 2)$-[β-D-Xylp-$(1\rightarrow 3)$]-β-D-Galp}	283
S. laxum	Luciamin, pale yellow powder, FAB-MS: 1085 $[M+Na]^+$, $[\alpha]_D^{20}$ −65.0° (c 0.3, MeOH)	(22R, 25S)-Spirost-5-ene-3β,15α-diol (**56**)	-3-O-{β-D-Glup-$(1\rightarrow 2)$-β-D-Glup-$(1\rightarrow 4)$-[α-L-Rhap-$(1\rightarrow 2)$]-β-D-Galp}	223
S. nigrum	Solanigroside C, WAP, $C_{51}H_{82}O_{26}$, ESI-MS: 1109 $[M-H]^-$, $[\alpha]_D^{25}$ −21.1° (c 0.54, MeOH)	(22R, 25R)-5α-Spirostane-3β,15α,23α-triol-26-one (**27**)	-3-O-β-D-Glup-$(1\rightarrow 2)$-O-[β-D-Glup-$(1\rightarrow 3)$]-O-β-D-Glup	188
	Solanigroside D, WAP, $C_{55}H_{88}O_{27}$, ESI-MS: 1203 $[M+Na]^+$, $[\alpha]_D^{25}$ −45.4° (c 0.84, MeOH)	(22R, 25R)-5α-Spirostane-3β,23α-diol-26-one (**17**)	-3-O-α-L-Arap-$(1\rightarrow 2)$-O-[β-D-Xylp-$(1\rightarrow 3)$]-O-β-D-Glup-$(1\rightarrow 4)$-O-[α-L-Rhap-$(1\rightarrow 2)$]-O-β-D-Galp	
	Solanigroside E, WAP, $C_{55}H_{88}O_{28}$, ESI-MS: 1219 $[M+Na]^+$, $[\alpha]_D^{25}$ −36.1° (c 1.07, MeOH)	(22R, 25R)-5α-Spirostane-3β,15α,23α-triol-26-one (**27**)	-3-O-α-L-Arap-$(1\rightarrow 2)$-O-[β-D-Xylp-$(1\rightarrow 3)$]-O-β-D-Glup-$(1\rightarrow 4)$-O-[α-L-Rhap-$(1\rightarrow 2)$]-O-β-D-Galp	

	Solanigroside F, WAP, $C_{56}H_{92}O_{28}$, ESI-MS: 1211 [M−H]⁻, $[\alpha]_D^{25}$ −37.6° (c 0.98, MeOH)	(25R)-5α-Spirostane-3β,23α-diol (**16**)	-3-O-β-D-Glup-$(1 \rightarrow 2)$-O-[β-D-Xylp-$(1 \rightarrow 3)$]-O-β-D-Glup-$(1 \rightarrow 4)$-O-β-D-Galp; -23-O-β-D-Glup	
	Solanigroside G, WAP, $C_{50}H_{82}O_{23}$, ESI-MS: 1049 [M−H]⁻, $[\alpha]_D^{25}$ −28.8° (c 0.41, MeOH)	(25R)-5α-Spirostane-3β,15α-diol (**14**)	3-O-β-D-Glup-$(1 \rightarrow 2)$-O-[β-D-Xylp-$(1 \rightarrow 3)$]-O-β-D-Glup-$(1 \rightarrow 4)$-O-β-D-Galp	
	Solanigroside H, WAP, $C_{51}H_{82}O_{22}$, ESI-MS: 1045 [M−H]⁻, $[\alpha]_D^{25}$ −63.9° (c 0.44, MeOH)	Pennogenin (**57**)	3-O-β-D-Glup-$(1 \rightarrow 2)$-O-α-L-Rhap-$(1 \rightarrow 4)$-O-[α-L-Rhap-$(1 \rightarrow 2)$]-O-β-D-Glup	
S. sisymbriifolium	Compound 1, $C_{45}H_{72}O_{18}$, 261–262°C, FAB-MS: 923 [M+Na]⁺, $[\alpha]_D^{20}$ −12.8° (c 0.08, EtOH)	Isonuatigenin (**61**)	-3-O-β-Solatrioside	284
S. sodomaeum	Compound 1, AP, $C_{51}H_{82}O_{21}$, FAB-MS: 1053.5393 [M+Na]⁺, $[\alpha]_D^{17}$ −97.8° (c 3.6, MeOH)	(25R, 26R)-26-Methoxy-spirost-5-en-3β-ol (**83**)	3-O-{O-α-L-Rhap-$(1 \rightarrow 2)$-O-[β-D-Xylp-$(1 \rightarrow 2)$-O-α-L-Rhap-$(1 \rightarrow 4)$]-β-D-Glup}	285
S. torvum	Torvoside J, AP, $C_{39}H_{64}O_{13}$, FAB-MS: 763 [M+Na]⁺, $[\alpha]_D$ −53.1° (c 0.4, MeOH)	(22R, 23S, 25S)-5α-Spirostane-3β,6α,23-triol (**24**)	6-O-α-L-Rhap-$(1 \rightarrow 3)$-β-D-Quinp	286
	Torvoside K, AP, $C_{39}H_{64}O_{13}$, FAB-MS: 763 [M+Na]⁺, $[\alpha]_D$ −59.3° (c 0.4, MeOH)	(22R, 23S, 25R)-5α-Spirostane-3β,6α,23-triol (**25**)	6-O-α-L-Rhap-$(1 \rightarrow 3)$-β-D-Quinp	

Table 1 (continued)

Plant name and family	Glycoside, physical nature, mp (°C), Mol. formula, Mol. wt. (m/z), $[\alpha]_D$	Aglycone/sapogenin	Sugar with linkage	Reference
	Torvoside L, AP, $C_{39}H_{64}O_{13}$, FAB-MS: 763 $[M+Na]^+$, $[\alpha]_D$ −3.8° (c 0.4, MeOH)	(22R, 23R, 25S)-5α-Spirostane-3β,6α,23-triol (**26**)	-6-O-α-L-Rhap-$(1 \rightarrow 3)$-β-D-Quinp	
	Torvoside H, AP, 170–172°, $C_{45}H_{73}O_{18}$, ESITOF-MS: 901.465 $[M-H]^-$, $[\alpha]_D^{29}$ −58.15° (c 0.114, MeOH)	(25S)-5α-Spiroxtane-6α,26-diol-3-one (**170**)	-6-O-α-L-Rhap-$(1 \rightarrow 3)$-β-D-Quinp; -26-O-β-D-Glup	76
Tacca chantrieri (Taccaceae)	Compound 1, AS, $C_{58}H_{96}O_{27}$, FAB-MS: 1223 $[M-H]^-$, $[\alpha]_D^{25}$ −82.0° (c 0.1, $CHCl_3$–MeOH)	(25S)-22α-Methoxy-furost-5-ene-3β,26-diol (**132**)	-3-O-α-L-Rhap-$(1 \rightarrow 2)$-O-[O-β-D-Glup-$(1 \rightarrow 4)$-α-L-Rhap-$(1 \rightarrow 3)$]-β-D-Glup; -26-O-β-D-Glup	137
	Compound 2, AS, $C_{60}H_{98}O_{28}$, FAB-MS: 1265 $[M-H]^-$, $[\alpha]_D^{25}$ −106.0° (c 0.1, $CHCl_3$–MeOH)	(25S)-22α-Methoxy-furost-5-ene-3β,26-diol (**132**)	-3-O-α-L-Rhap-$(1 \rightarrow 2)$-O-[O-β-D-Glup-$(1 \rightarrow 4)$-α-L-Rhap-$(1 \rightarrow 3)$]-6-O-acetyl-β-D-Glup; -26-O-β-D-Glup	
	Compound 3, AS, $C_{64}H_{106}O_{32}$, FAB-MS: 1385 $[M-H]^-$, $[\alpha]_D^{25}$ −54.0° (c 0.1, $CHCl_3$–MeOH)	(25S)-22α-Methoxy-furost-5-ene-3β,26-diol (**132**)	-3-O-α-L-Rhap-$(1 \rightarrow 2)$-O-[O-β-D-Glup-$(1 \rightarrow 4)$-α-L-Rhap-$(1 \rightarrow 3)$-β-D-Glup; -26-O-β-D-Glup-$(1 \rightarrow 6)$-β-D-Glup	

Compound 4, AS, $C_{57}H_{92}O_{26}$, FAB-MS: 1215 [M+Na]$^+$, $[\alpha]_D^{25}$ −60.0° (c 0.10, CHCl$_3$–MeOH)	(25S)-Furosta-5,20(22)-diene-3β,26-diol (**143**)	-3-O-α-L-Rhap-(1→2)-O-[O-β-D-Glup-(1→4)-α-L-Rhap-(1→3)]-β-D-Glup; -26-O-β-D-Glup	
Compound 5, AS, $C_{59}H_{94}O_{27}$, HR-MALDITOFMS: 1257.5891 [M+Na]$^+$, $[\alpha]_D^{25}$ −42.0° (c 0.1, CHCl$_3$–MeOH)	(25S)-Furosta-5,20(22)-diene-3β,26-diol (**143**)	-3-O-α-L-Rhap-(1→2)-O-[O-β-D-Glup-(1→4)-α-L-Rhap-(1→3)]-6-O-acetyl-β-D-Glup; -26-O-β-D-Glup	
Compound 1, AS, $C_{51}H_{82}O_{21}$, HR-FAB-MS: 1053.5208 [M+Na]$^+$, $[\alpha]_D^{25}$ −86.0° (c 0.1, CHCl$_3$–MeOH, 1:1)	Yamogenin (**50**)	-3-O-α-L-Rhap-(1→2)-O-[O-β-D-Glup-(1→4)-α-L-Rhap-(1→3)]-β-D-Glup	*182*
Compound 2, AS, $C_{51}H_{82}O_{22}$, HR-FAB-MS: 1069.5195 [M+Na]$^+$, $[\alpha]_D^{25}$ −108.0° (c 0.1, CHCl$_3$–MeOH, 1:1)	(24S, 25R)-Spirost-5-ene-3β,24-diol (**60**)	-3-O-α-L-Rhap-(1→2)-O-[O-β-D-Glup-(1→4)-α-L-Rhap-(1→3)]-β-D-Glup	

Table 1 (continued)

Plant name and family	Glycoside, physical nature, mp (°C), Mol. formula, Mol. wt. (m/z), $[\alpha]_D$	Aglycone/sapogenin	Sugar with linkage	Reference
	Compound 3, AS, $C_{45}H_{72}O_{17}$, HR-FAB-MS: 907.4692 $[M+Na]^+$, $[\alpha]_D^{25} -86.0°$ (c 0.1, CHCl$_3$–MeOH, 1:1)	Yamogenin (**50**)	-3-O-β-D-Glup- -(1→4)-O-α-L-Rhap-(1→3)- β-D-Glup	
	Compound 4, AS, $C_{45}H_{72}O_{17}$, HR-FAB-MS: 885.4810 $[M+H]^+$, $[\alpha]_D^{25} -112.0°$ (c 0.1, CHCl$_3$–MeOH, 1:1)	(24S, 25R)-Spirost-5-ene-3β,24-diol (**60**)	-3-O-α-L-Rhap- (1→2)-O-[α-L-Rhap- (1→3)]-β-D-Glup	
Tribulus alatus (Zygophyllaceae)	Compound 1, WAP, $C_{57}H_{96}O_{29}$, ESI-MS: 1267 $[M+Na]^+$, $[\alpha]_D^{25} -61.0°$ (c 0.1, MeOH)	(25S)-5α-Furostane-3β,22α,26-triol (**99**)	-3-O-β-D-Galp- (1→2)-O-[β-D-Glup- (1→3)]-O-β-D-Glup- (1→4)-β-D-Galp; -26-O-β-D-Glup	287
	Compound 2, WAP, $C_{57}H_{96}O_{29}$, ESI-MS: 1243 $[M-H]^-$, $[\alpha]_D^{25} -94.0°$ (c 0.1, MeOH)	(25S)-5α-Furostane-3β,22α,26-triol (**99**)	-3-O-β-D-Glup- (1→2)-O-[β-D-Glup- (1→3)]-O-β-D-Glup- (1→4)-β-D-Galp; -26-O-β-D-Glup	
	Compound 4, WAP, $C_{51}H_{84}O_{24}$, ESI-MS: 1103 $[M+Na]^+$, $[\alpha]_D^{25} -45.0°$ (c 0.1, MeOH)	Neogitogenin (**5**)	-3-O-β-D-Galp- (1→2)-O-[β-D-Glup- (1→3)]-O-β-D-Glup- (1→4)-β-D-Galp	

	Compound 5, WAP, C$_{51}$H$_{84}$O$_{23}$; ESI-MS: 1087 [M + Na]$^+$, [α]$_D^{25}$ −23.0° (c 0.1, MeOH)	Neotigogenin (**2**)	-3-*O*-β-D-Galp- (1 → 2)-*O*-[β-D-Glup- (1 → 3)]-*O*-β-D-Glup- (1 → 4)-β-D-Galp	288
T. parvispinus	Parvispinoside A, AP, C$_{56}$H$_{94}$O$_{29}$; ESI-MS: 1253 [M + Na]$^+$, [α]$_D^{22}$ +7.1° (c 0.1, MeOH)	(25*R*)-5α-Furostane-2α,3β,22α,26-tetrol (**102**)	-3-*O*-{β-D-Galp- (1 → 2)-*O*-[β-D-Xylp- (1 → 3)]-*O*-β-D-Glup- (1 → 4)-β-D-Galp}; -26-*O*-β-D-Glup	
	Parvispinoside B, AP, C$_{56}$H$_{94}$O$_{28}$; HR-MALDI-MS: 1237.5844 [M + Na]$^+$, [α]$_D^{22}$ −29.1° (c 0.1, MeOH)	(25*R*)-5α-Furostane-3β,22α,26-triol (**98**)	-3-*O*-{β-D-Galp- (1 → 2)-*O*-[β-D-Xylp- (1 → 3)]-*O*-β-D-Glup- (1 → 4)-β-D-Galp}; -26-*O*-β-D-Glup	
	22-*O*-Methyl-parvispinoside A, AP, C$_{57}$H$_{96}$O$_{29}$, HR-MALDI-MS: 1267.5946 [M + Na]$^+$, [α]$_D^{22}$ −11.9° (c 1.9, MeOH)	(25*R*)-22α-Methoxy-5α-furostane-2α,3β,26-triol (**95**)	-3-*O*-{β-D-Galp- (1 → 2)-*O*-[β-D-Xylp- (1 → 3)]-*O*-β-D-Glup- (1 → 4)-β-D-Galp}; -26-*O*-β-D-Glup	
	22-*O*-Methyl-parvispinoside B, AP, C$_{57}$H$_{96}$O$_{28}$, HR-MALDI-MS: 1251.5998 [M + Na]$^+$, [α]$_D^{22}$ −14.3° (c 0.1, MeOH)	(25*R*)-22α-Methoxy-5α-furostane-3β,26-diol (**94**)	-3-*O*-{β-D-Galp- (1 → 2)-*O*-[β-D-Xylp- (1 → 3)]-*O*-β-D-Glup- (1 → 4)-β-D-Galp}; -26-*O*-β-D-Glup	
T. terrestris	Neoprotodioscin, AP C$_{51}$H$_{86}$O$_{22}$, 207–208°C, ESI-MS: 1049 [M−H]$^-$	(25*R*)-5α-Furostane-3β,22α,26-triol (**98**)	-3-*O*-α-L-Rhap- (1 → 2)-*O*-[α-L-Rhap- (1 → 4)]-β-D-Glup	289

Table 1 (continued)

Plant name and family	Glycoside, physical nature, mp (°C), Mol. formula, Mol. wt. (m/z), $[\alpha]_D$	Aglycone/sapogenin	Sugar with linkage	Reference
	Tribulosaponin A, WP, $C_{51}H_{84}O_{21}$, HR-ESIFT-MS: 1033.5636 $[M]^+$, $[\alpha]_D^{25}$ −73.0° (c 0.004, MeOH)	(25S)-5β-Furost-20(22)-ene-3β,26-diol (**147**)	-3-O-α-L-Rhap-$(1\rightarrow 2)$-[α-L-Rhap-$(1\rightarrow 4)$]-β-D-Glup; -26-O-β-D-Glup	*136*
	Tribulosaponin B, WP, $C_{51}H_{84}O_{22}$, HR-ESIFT-MS: 1049.5611 $[M]^+$, $[\alpha]_D^{25}$ −34.0° (c 0.004, MeOH)	(25S)-5β-Furost-20(22)-ene-3β,26-diol (**147**)	-3-O-α-L-Rhap-$(1\rightarrow 2)$-[β-D-Glup-$(1\rightarrow 4)$]-β-D-Galp; -26-O-β-D-Glup	
	Isoterrestrosin B, WP, $C_{45}H_{74}O_{17}$, HR-ESIFT-MS: 887.4903 $[M+H]^+$, $[\alpha]_D^{25}$ −140.0° (c 0.004, MeOH)	Sarsasapogenin (**34**)	-3-O-α-L-Rhap-$(1\rightarrow 2)$-[β-D-Glup-$(1\rightarrow 4)$]-β-D-Galp	*290*
	Compound 1	Hecogenin (**32**)	-3-O-β-D-Xylp-$(1\rightarrow 3)$-β-D-Glup-$(1\rightarrow 4)$-β-D-Galp	
	Compound 2	Hecogenin (**32**)	-3-O-β-D-Glup-$(1\rightarrow 2)$-β-D-Glup-$(1\rightarrow 4)$-β-D-Galp	
	Compound 3	(25R)-22α-Methoxy-5α-furostane-3β,26-diol (**94**)	-3-O-{β-D-Xylp-$(1\rightarrow 2)$-[β-D-Xylp-$(1\rightarrow 3)$]-β-D-Glup-$(1\rightarrow 4)$-[α-L-Rhap-$(1\rightarrow 2)$]-β-D-Galp}; -26-O-β-D-Glup	

	Compound	Sugars	Ref.
	Methyl prototribestin, AP, $C_{46}H_{75}O_{21}SNa$, ESI-MS: 1041 [M + Na]$^+$, (25R)-22α-Methoxy-furost-5-ene-3β,26-diol (**131**)	-3-O-α-L-Rhap-(1→2)-β-D-{4-O-sulpho}-Glup; -26-O-β-D-Glup	291
	Prototribestin, AP, $C_{45}H_{73}O_{21}SNa$, ESI-MS: 1027 [M + Na]$^+$ (22α, 25R)-Furost-5-ene-3β,22,26-triol (**127**)	-3-O-α-L-Rhap-(1→2)-β-D-{4-O-sulpho}-Glup; -26-O-β-D-Glup	
	Terrestrinin A (25S)-Furost-4,20(22)-diene-26-ol-3,12-dione (**171**)	-26-O-β-D-Glup	292
	Terrestrinin B (25S)-5α-Furostane-3β,22α,26-triol (**99**)	-3-O-β-D-Xylp-(1→3)-[β-D-Xylp-(1→2)]-β-D-Glup-(1→4)-[α-L-Rhap-(1→2)]-β-D-Galp; -26-O-β-D-Glup	
Trigonella foenum-graecum (Leguminosae)	SA III, colorless needles, 242–244°C, $C_{38}H_{60}O_{12}$, [M]$^+$, 708, $[\alpha]_D^{20}$ −96.0° (CHCl$_3$) Yamogenin (**50**)	-3-O-β-D-Glup-(1→4)-α-D-Xylp;	293
	Trigoneoside Xa (25S)-5α-Furostane-2α,3β,22ξ,26-tetrol (**104**)	-3-O-α-L-Rhap-(1→2)-β-D-Glup; -26-O-β-D-Glup	294
	Trigoneoside Xb (25R)-5α-Furostane-2α,3β,22ξ,26-tetrol (**103**)	-3-O-α-L-Rhap-(1→2)-β-D-Glup; -26-O-β-D-Glup	

Table 1 (continued)

Plant name and family	Glycoside, physical nature, mp (°C), Mol. formula, Mol. wt. (m/z), $[\alpha]_D$	Aglycone/sapogenin	Sugar with linkage	Reference
	Trigoneoside XIb	(25R)-5α-Furostane-2α,3β,22ξ,26-tetrol (**103**)	-3-O-β-D-Xylp-$(1 \rightarrow 4)$-β-D-Glup; -26-O-β-D-Glup	
	Trigoneoside XIIa	(25S)-Furost-4-ene-3β,22ξ,26-triol (**173**)	-3-O-α-L-Rhap-$(1 \rightarrow 2)$-β-D-Glup; -26-O-β-D-Glup	
	Trigoneoside XIIb	(25R)-Furost-4-ene-3β,22ξ,26-triol (**172**)	-3-O-α-L-Rhap-$(1 \rightarrow 2)$-β-D-Glup; -26-O-β-D-Glup	
	Trigoneoside XIIIa	(25S)-Furost-5-ene-3β,22ξ,26-triol (**128**)	-3-O-α-L-Rhap-$(1 \rightarrow 2)$-[β-D-Glup-$(1 \rightarrow 3)$-β-D-Glup-$(1 \rightarrow 4)$]-β-D-Glup; -26-O-β-D-Glup	
Trillium kamtschaticum (Trilliaceae)	Trillenoside C, AP, FAB-MS: 777 $[M+Na]^+$, $[\alpha]_D^9$ −120.0° (c 1.8, MeOH)	Trillenogenin (**167**)	-1-O-α-L-Rhap-$(1 \rightarrow 2)$-α-L-Arap	49
	Deoxytrillenoside B, AP, FAB-MS: 871 $[M+H]^+$, $[\alpha]_D^{25}$ −92.1° (c 0.80, MeOH)	21-Deoxytrillenogenin (**168**)	-1-O-α-L-Rhap-$(1 \rightarrow 2)$-[β-D-Xylp-$(1 \rightarrow 3)$]-α-L-Arap	
Tupistra wattii (Convallariaceae)	Wattoside G, AP, 214–216°C, $C_{32}H_{52}O_{11}$, HR-FAB-MS: 611.3466 $[M-H]^-$, $[\alpha]_D^{20}$ −65.5° (c 0.03, MeOH)	Pentologenin (**41**)	-4-O-β-D-Xylp	295

	Wattoside H, Colorless needles, $C_{33}H_{52}O_{15}$, 200–203°C, HR-FAB-MS: 687.3278 $[M-H]^-$, $[\alpha]_D^{20}$ −78.0° (c 0.014, MeOH)	(24S, 25S)-Spirostane-1β,2β,3β,4β,5β,7β,24-heptol-6-one (**42**)	-24-O-β-D-Glup	
	Wattoside I, AP, $C_{39}H_{64}O_{15}$, 205–207°C, HR-FAB-MS: 771.4153 $[M-H]^-$, $[\alpha]_D^{20}$ −76.2° (c 0.027, MeOH)	(24S, 25S)-5β-Spirostane-1β,3β,24-triol (**38**)	-24-O-β-D-Glup-(1→6)-β-D-Glup	
T. yunnanensis	Tupistroside A, WAP, $C_{32}H_{52}O_{10}$, FAB-MS: 595 $[M-H]^-$, $[\alpha]_D$ −56.4° (c 0.4, MeOH)	Convallogenin B (**39**)	-3-O-α-L-Arap	296
	Tupistroside B, WAP, $C_{33}H_{50}O_{10}$, HR-FAB-MS: 605.3326 $[M-H]^-$, $[\alpha]_D$ −60.5° (c 0.4, MeOH)	Spirost-5,25(27)-diene-1β,3α,24β-triol (**88**)	-3-O-β-D-Glup	
	Tupistroside C, WAP, $C_{33}H_{54}O_{12}$, HR-FAB-MS: 641.3537 $[M-H]^-$, $[\alpha]_D$ −67.2° (c 0.40, MeOH)	(22S, 25S)-Furospirostane-1α,2β,3α,5α,26-pentol (**157**)	-26-O-β-D-Glup	
	Tupistroside D, WAP, $C_{33}H_{52}O_{10}$, HR-FAB-MS: 607.3436 $[M-H]^-$, $[\alpha]_D$ −50.6° (c 0.3, Pyr)	Furost-5,25(27)-diene-1β,3α,22ξ,26-tetrol (**124**)	-26-O-β-D-Glup	
	Tupistroside E, WAP, $C_{39}H_{64}O_{15}$, HR-FAB-MS: 771.4150 $[M-H]^-$, $[\alpha]_D$ −48.5° (c 0.3, Pyr)	Furost-5-ene-1β,3α,22,26-tetrol (**142**)	-3-O-β-D-Glup; -26-O-β-D-Glup	
	Tupistroside F, WAP, $C_{34}H_{54}O_{15}$, HR-FAB-MS: 701.3386 $[M-H]^-$, $[\alpha]_D$ −51.7° (c 0.2, Pyr)	22ξ-Methoxy-furost-25(27)-ene-1β,2β,3β,4β,5β,7α,26-heptol-6-one (**122**)	-26-O-β-D-Glup	

Table 1 (continued)

Plant name and family	Glycoside, physical nature, mp (°C), Mol. formula, Mol. wt. (m/z), $[\alpha]_D$	Aglycone/sapogenin	Sugar with linkage	Reference
Veronica fushii and *V. multifida* (Scrophulariaceae)	Aculeatiside A, WAP, $C_{51}H_{82}O_{22}$, HR-FAB-MS: 1069.5 $[M+Na]^+$	Nuatigenin (**154**)	-3-O-[α-L-Rhap-$(1\rightarrow 2)$-α-L-Rhap-$(1\rightarrow 4)$-β-D-Glup]; -26-O-β-D-Glup	297
	Mulifidoside, WAP, $C_{57}H_{92}O_{27}$, HR-FAB-MS: 1231.5 $[M+Na]^+$, $[\alpha]_D^{20}$ −78.0°	Nuatigenin (**154**)	-3-O-{[α-L-Rhap-$(1\rightarrow 2)$]-[β-D-Glup-$(1\rightarrow 4)$-α-L-Rhap-$(1\rightarrow 4)$]-β-D-Glup}; -26-O-β-D-Glup	
Yucca filamentosa (Agavaceae)	Compound 1, $C_{63}H_{104}O_{33}$, Colorless needles, 292–293°C, ESI-MS: 1412 $[M+H+Na]^+$	Gitogenin (**4**)	-3-O-{[β-D-Glup-$(1\rightarrow 3)$-β-D-Glup-$(1\rightarrow 2)$]-{α-L-Rhap-$(1\rightarrow 4)$-β-D-Glup-$(1\rightarrow 3)$]-β-D-Glup-$(1\rightarrow 4)$]-β-D-Galp]	85
Y. schidigera	Compound 5, WP, $C_{45}H_{75}O_{19}$, 207–208°C, HR-MS: 919.4917 $[M]^-$, $[\alpha]_D^{25}$ −38.8° (c 0.1, MeOH)	(25R)-5β-Furostane-3β,22α,26-triol (**114**)	-3-O-β-D-Glup-$(1\rightarrow 2)$-β-D-Glup; 26-O-β-D-Glup	298
	Compound 6, WP, $C_{50}H_{81}O_{23}$, 235–236°C, HR-MS: 1049.5166 $[M]^-$, $[\alpha]_D^{25}$ −43.25° (c 0.1, MeOH)	(25R)-5β-Furostane-3β,22α,26-triol (**114**)	-3-O-β-D-Glup-$(1\rightarrow 2)$-[β-D-Xylp-$(1\rightarrow 3)$]-β-D-Glup; -26-O-β-D-Glup	

Compound 7, WP, $C_{50}H_{79}O_{23}$, 193–195°C, HR-MS: 1047.5002 [M]$^-$, $[\alpha]_D^{25}$ −3.6° (c 0.1, MeOH)	(25R)-5β-Furost-20(22)-ene-3β,26-diol-12-one (**148**)	-3-O-β-D-Glup-(1→2)-[β-D-Xylp-(1→3)]-β-D-Glup; -26-O-β-D-Glup
Schidegera saponin A1, WAP, $C_{44}H_{69}O_{17}$, FAB-MS: 869 [M−H]$^-$, $[\alpha]_D^{24}$ −44.6° (c 1.11, MeOH)	Macranthogenin (**44**)	-3-O-β-D-Xylp-(1→3)-[β-D-Glup-(1→2)]-β-D-Glup
Schidegera saponin A2, WAP, $C_{44}H_{69}O_{17}$, FAB-MS: 869 [M−H]$^-$, $[\alpha]_D^{28}$ −55.2° (c 0.52, Pyr)	Macranthogenin (**44**)	-3-O-β-D-Xylp-(1→3)-[β-D-Glup-(1→2)]-β-D-Galp
Schidegera saponin A3, WAP, $C_{44}H_{71}O_{18}$, FAB-MS: 899 [M−H]$^-$, $[\alpha]_D^{24}$ −52.2° (c 1.71, MeOH)	Macranthogenin (**44**)	-3-O-β-D-Glup-(1→3)-[β-D-Glup-(1→2)]-β-D-Glup
Schidegera saponin B1, WAP, $C_{44}H_{67}O_{18}$, FAB-MS: 883 [M−H]$^-$, $[\alpha]_D^{24}$ −10.3° (c 1.71, MeOH)	5β-Spirost-25(27)-en-3β-ol-12-one (**45**)	-3-O-β-D-Glup-(1→3)-[β-D-Glup-(1→2)]-β-D-Glup
Schidegera saponin C1, WAP, $C_{44}H_{69}O_{18}$, FAB-MS: 885 [M−H]$^-$, $[\alpha]_D^{24}$ −, 56.4° (c 0.11, MeOH)	Schidegeragenin C (**46**)	-3-O-β-D-Xylp-(1→3)-[β-D-Glup-(1→2)]-β-D-Galp
Schidegera saponin C2, WAP, $C_{39}H_{61}O_{14}$, FAB-MS: 753 [M−H]$^-$, $[\alpha]_D^{24}$ −38.2° (c 0.55, MeOH)	Schidegeragenin C (**46**)	-3-O-β-D-Glup-(1→2)-β-D-Galp

Mol., Molecular; WP, white powder; AP, amorphous powder; WAS, white amorphous solid; WAP, white amorphous powder; AS, amorphous solid; Pyr, pyridine; Glu, glucose; Gal, galactose; Gul, gulose; Rha, rhamnose; Fuc, fucose; Apio, apiose; Xyl, xylose; Ara, arabinose; Qui, quinovose; p, pyranosyl; f, furanosyl.

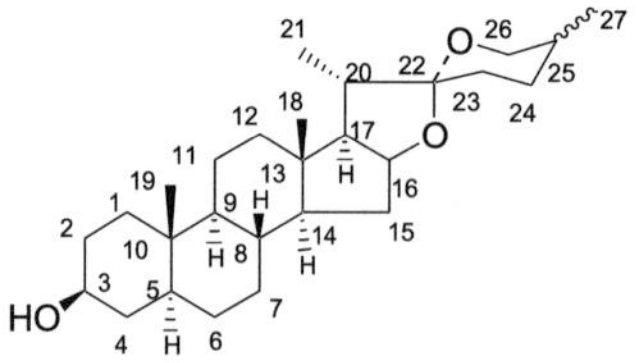

39. 1β,4β,5β-OH (25S): Convallogenin B
40. 1β,2β,3α,24(S)-OH (25R):
    (24S)-Hydroxy-neotokorogenin
41. 1β,2β,4β,5β-OH (25R): Pentologenin
42. 1β,2β,4β,5β,7β,24(S)-OH,6-oxo (25S)
43. 12-oxo (25R): Gloriogenin

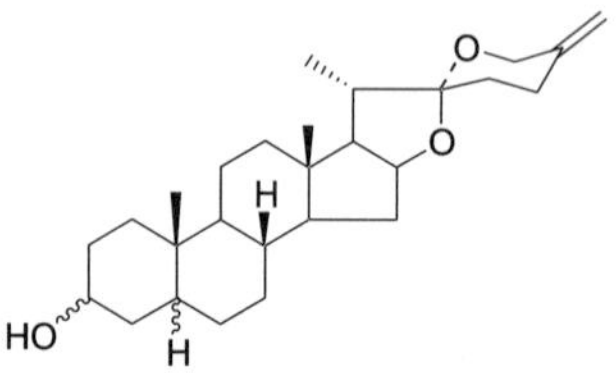

1. (25R): Tigogenin
2. (25S): Neotigogenin
3. 1β-OH (25R)
4. 2α-OH (25R): Gitogenin
5. 2α-OH (25S): Neogitogenin
6. 2α-OH 12-oxo (25R): Manogenin
7. 2α-OH, 9(11)ene, 12-oxo:
   9,11-Dehydromanogenin
8. 6α-OH (25R): Chlorogenin
9. 6α-OH (25S): Neochlorogenin
10. 6α-OH, 12-oxo (25R)
11. 6β-OH (25R): β-Chlorogenin
12. 6β-OH, 2-oxo: Porrigenin B
13. 12β-OH (25R): Rockogenin
14. 15α-OH (25R)
15. 15α-OH,12-oxo (25R)
16. 23-OH (25R)
17. 23-OH,26-oxo (22R,25R)
18. 1β,2α-OH (25R)
19. 2α,6α-OH (25R)
20. 2α,6β-OH: Agigenin
21. 2α,12β-OH (25R)
22. 2α-OH, 27-CH2OH (25S): Crestagenin
23. 6α,23α-OH: Chrysogenin
24. 6α,23-OH (22R,23S,25S)
25. 6α,23-OH (22R,23S,25R)
26. 6α,23-OH (22R,23R,25S)
27. 15α,23α-OH,26-oxo (22R,25R)
28. 23,24-OH (25S)
29. 6α,23,24-OH (25S): Agavegenin C
30. 2α,5α,6β,24-OH (24S,25S)
31. 12-oxo (25S): Neohecogenin
32. 12-oxo (25R): Hecogenin
33. 12-oxo, 9(11)-ene (25R)

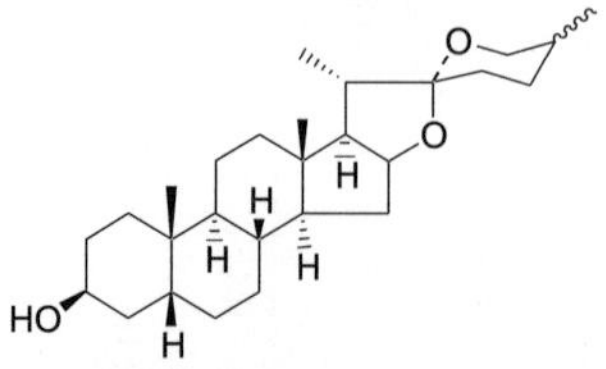

44. 5β H,3β-OH: Macranthogenin
45. 5β H,3β-OH, 12-oxo
46. 5β H,2β,3β-OH: Schidegeragenin C
47. 5α H,1β,3α-OH: 1β-Hydroxycrabbogenin
48. 5α H,1β,2α,3β-OH

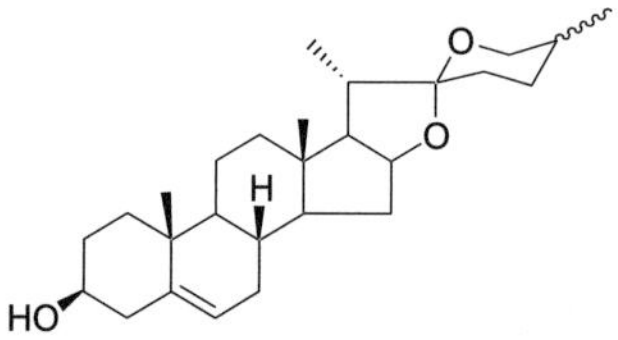

49. (22R,25R): Diosgenin
50. (22R,25S): Yamogenin
51. (22S,25R): Epiyamogenin
52. 1β-OH (25R): Ruscogenin
53. 1β-OAc (25R): Ruscogenin 1-acetate
54. 2α-OH (25R): Yuccagenin
55. 7α-OH (25R)
56. 15α-OH (22R,25S)
57. 17α-OH, (25R): Pennogenin
58. 17α-OH, 26(R)-OMe (25R)
59. 23(S)-OH (25R)
60. 24(S)-OH (25R)
61. 25-OH: Isonuatigenin
62. 26(R)-OH (25R)
63. 27-CH₂OH (25S): Isonarthogenin
64. 27-CH₂OH, 12-oxo (25S)
65. 1β,2α-OH (25R)
66. 1β,24(S)-OH (25S)
67. 1β,24(S)-OH (25R)
68. 2α,15β-OH (25R)
69. 2α,17α-OH (25R)
70. 2α,24-OH (24S,25R)
71. 14α,17α-OH (25R): Ophiojaponin C
72. 14,24-OH
73. 14,27-OH

34. (25S): Sarsasapogenin
35. (25R): Smilagenin
36. 1β-OH (25R): Isorhodeasapogenin
37. 17α-OH (25S)
38. 1β,24(S)-OH (25S)

References, pp. 127–141

74. 17α,23(S)-OH (25S)
75. 23(S),24(R)-OH (25S)
76. 1β-OAc,23(S),24(R)-OH (25R)
77. 23(S),26(R)-OH (22R,25R)
78. 23(S),27-OH,12-oxo (25S)
79. 1β,2α,17α-OH
80. 1β,23(S),24(S)-OH,15-oxo
81. 12α,17α,23(S)-OH (25R)
82. 14α,17α,23(S)-OH (25R)
83. 26(R)-OMe (25R)

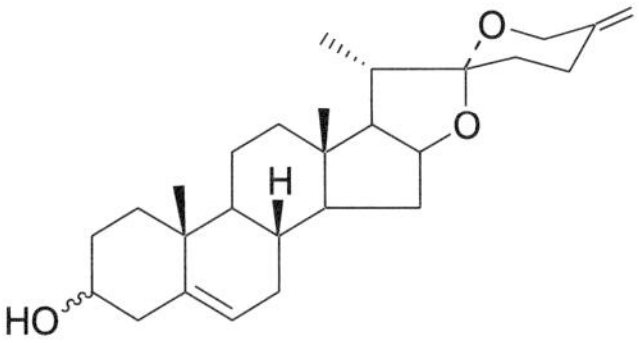

84. 3β-OH Sceptrumgenin
85. 1β,3β-OH Neoruscogenin
86. 1β,2α,3β-OH
87. 1β,3β,23(S)-OH
88. 1β,3α,24β-OH
89. 1β,2α,3β,12β-OH
90. 1β,2α,3β,23α-OH
91. 1β,3β,23(S),24(S)-OH
92. 1β,3β,23(S),24(S)-OH, 21-OAc

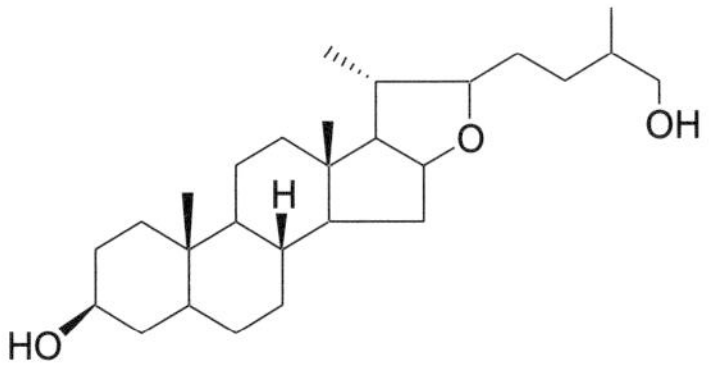

93. 5α H, 22ξ-OMe (25R)
94. 5α H, 22α-OMe (25R)
95. 5α H, 2α-OH, 22α-OMe (25R)
96. 5α H, 2α-OH, 22ξ-OMe,12-oxo (25R)
97. 5α H, 2α-OH, 22ξ-OMe,12-oxo, 9(11)-ene (25R)
98. 5α H, 22α-OH (25R)
99. 5α H, 22α-OH (25S)
100. 5α H, 22ξ-OH (25R)
101. 5α H, 22α-OH 12-oxo (25R)
102. 5α H, 2α,22α-OH (25R)
103. 5α H, 2α,22ξ-OH (25R)
104. 5α H, 2α,22ξ-OH (25S)
105. 5α H, 6β,22ξ-OH (25R)
106. 2α,5α,22α-OH
107. 2α,5α,22β-OH
108. 2α,5α,6β-OH, 22ξ-OMe (25R)

109. 2α,5α,6β,22α-OH
110. 2α,5α,6β,22β-OH
111. 5β H, 22α-OMe (25S)
112. 5β H, 22α-OMe (25R)
113. 5β H, 22(R)-OH (25R)
114. 5β H, 22α-OH (25R)
115. 5β H, 22-OH (25S)
116. 5β,6α,22ξ-OH (25S)

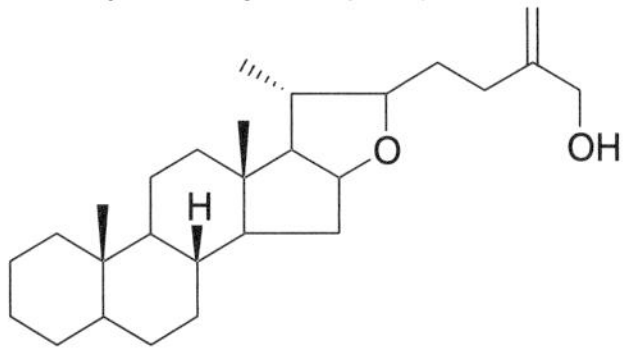

117. 5α H, 1β,3α-OH, 22ξ-OMe
118. 5α H, 1β,3β-OH, 22ξ-OMe
119. 5α H, 1β,3α,4α-OH, 22ξ-OMe
120. 5α H, 1β,3β,4α-OH, 22ξ-OMe
121. 5β H, 1β,3β,6β,22α-OH
122. 5β H, 1β,2β,3β,4β,5β,7α-OH, 22ξ-OMe, 6-oxo

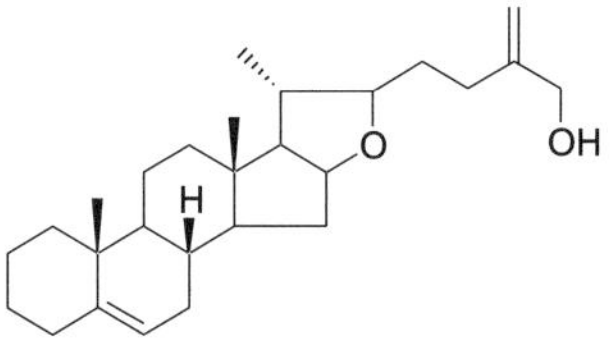

123. 1β,3β-OH, 22ξ-OMe
124. 1β,3α,22ξ-OH
125. 1β,2α,3β-OH, 22ξ-OMe
126. 1β,2α,3β,22ξ-OH

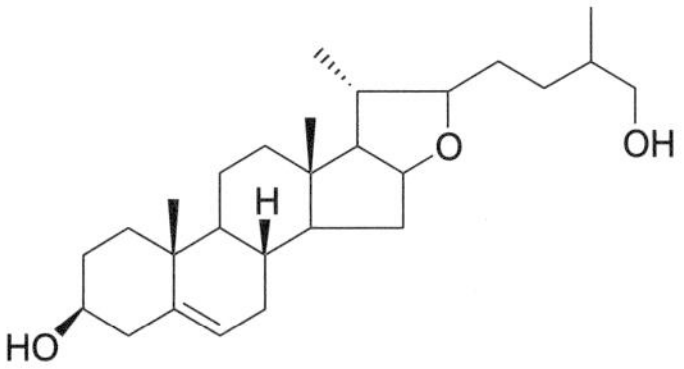

127. 22α-OH (25R)
128. 22ξ-OH (25S)
129. 22ξ-OH, 12-oxo (25S)
130. 22ξ-OH, 12-oxo (25R)
131. 22α-OMe (25R)
132. 22α-OMe (25S)
133. 22ξ-OMe (25R)
134. 22ξ-OMe (25S)
135. 22(R)-OMe (25R)
136. 22(R)-OMe (25S)
137. 1β,22ξ-OH (25R)

138. 1β,22ξ-OH (25S)
139. 1β-OH, 22ξ-OMe (25R)
140. 1β-OH, 22α-OMe (25S)
141. 2α-OH, 22α-OMe (25R)

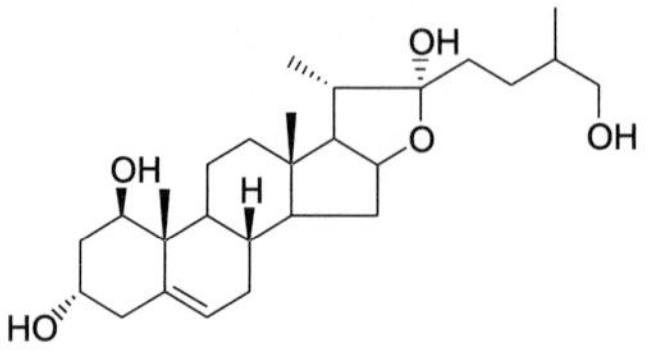

142

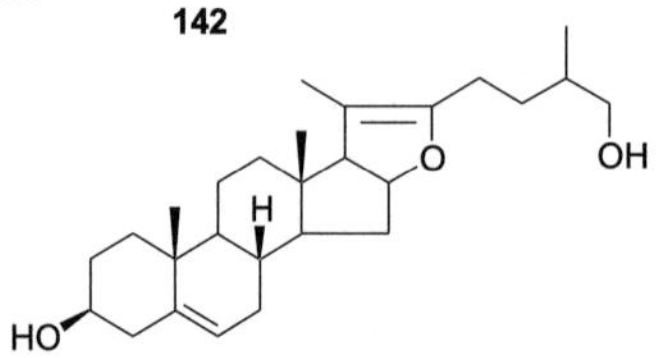

143. (25S)
144. 23(S)-OMe (25R)
145. 1β-OH (25R)
146. 2α-OH (25R)

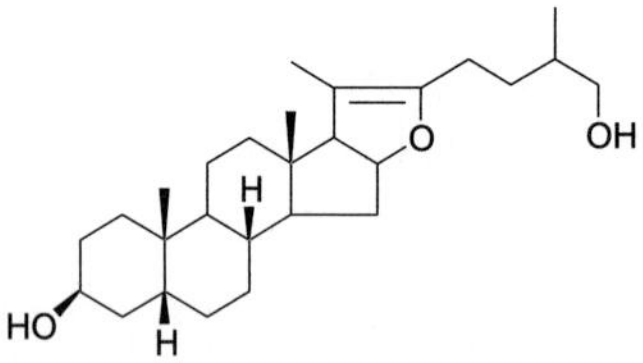

147. (25S)
148. 12-oxo (25R)

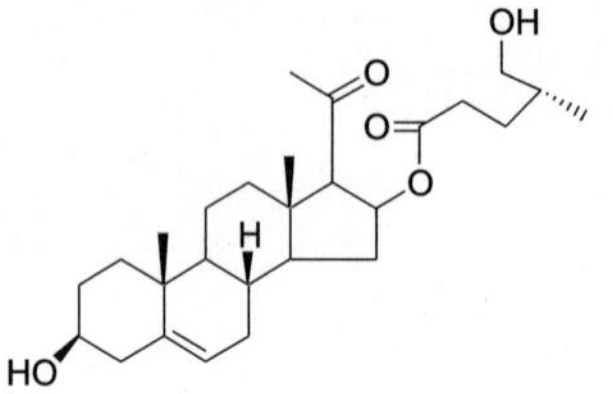

149. (25R)
150. 23(S)-OH (25R)

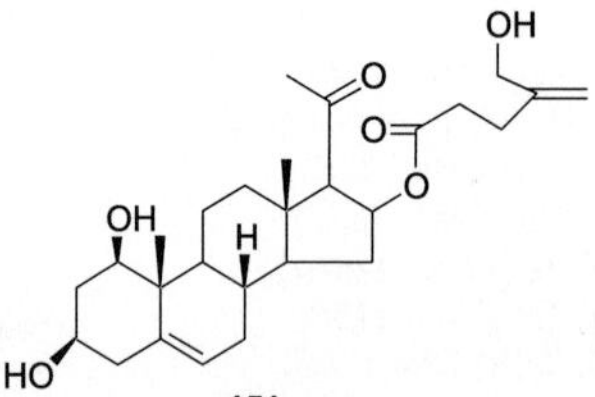

151

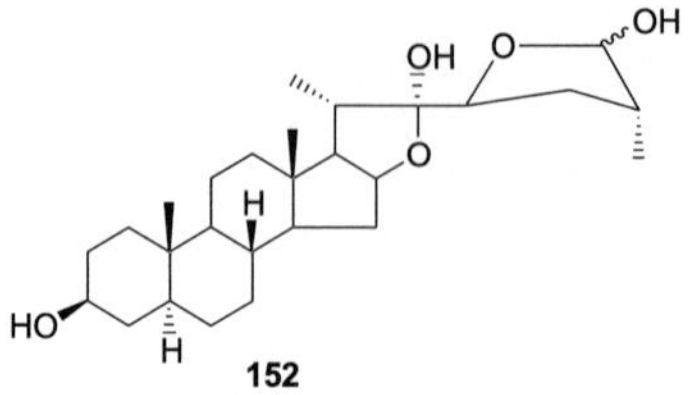

152

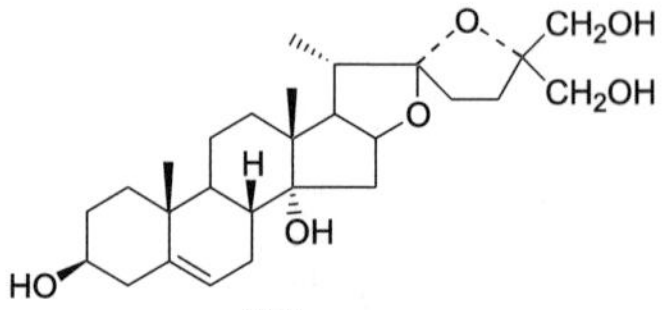

153

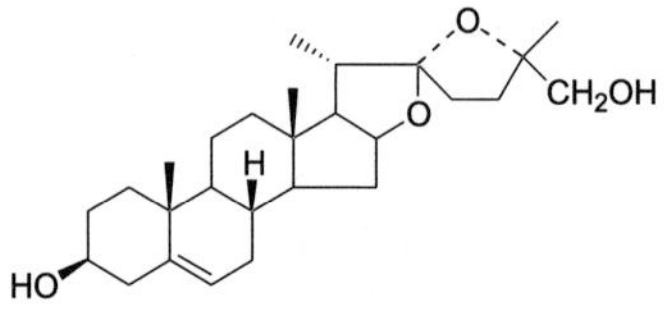

154. (25S): Nuatigenin
155. 7β-OH
156. 7β-OMe

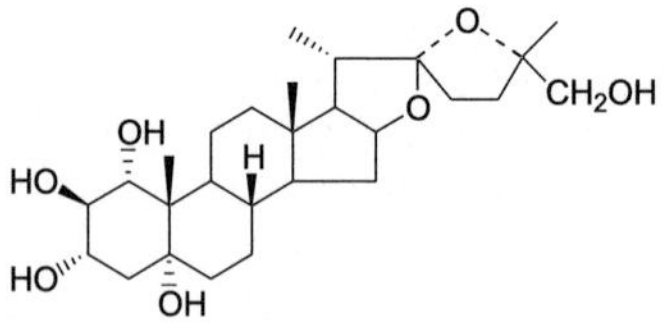

157

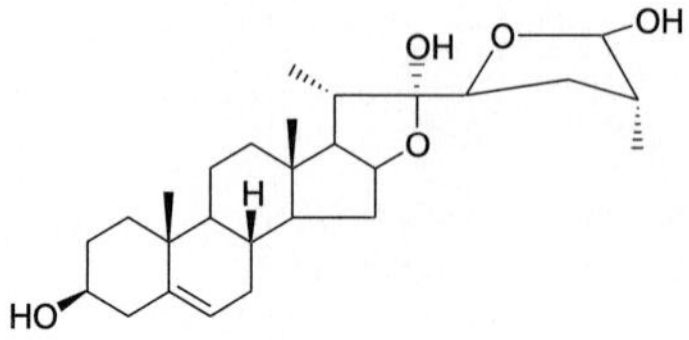

158

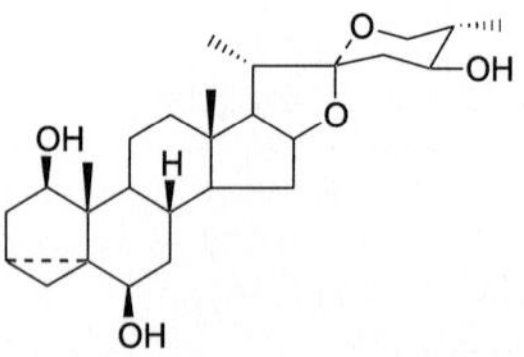

159

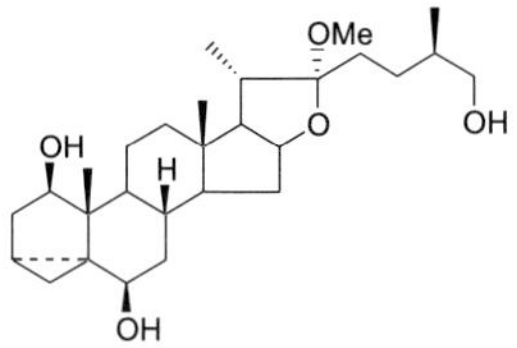

160

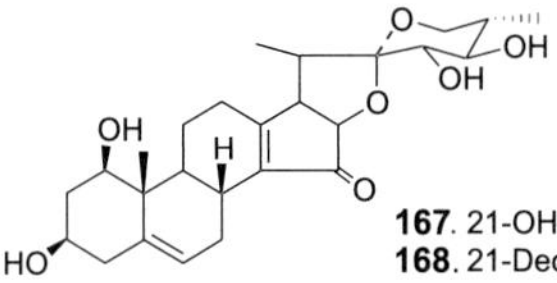

161

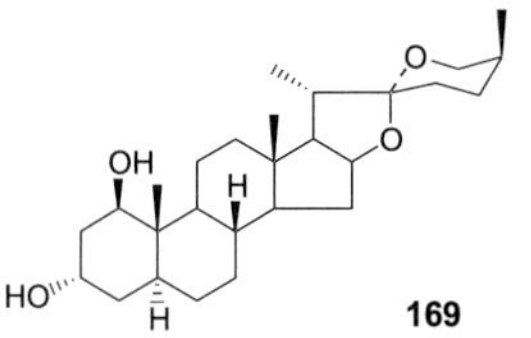

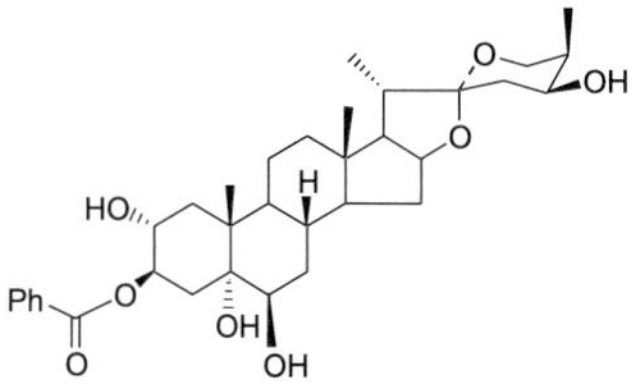

162

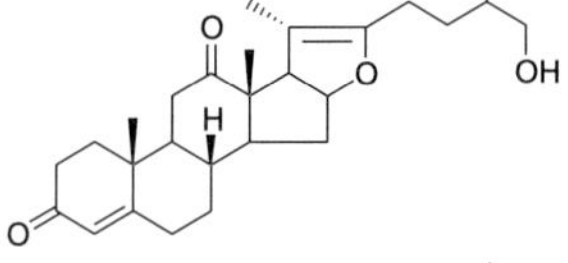

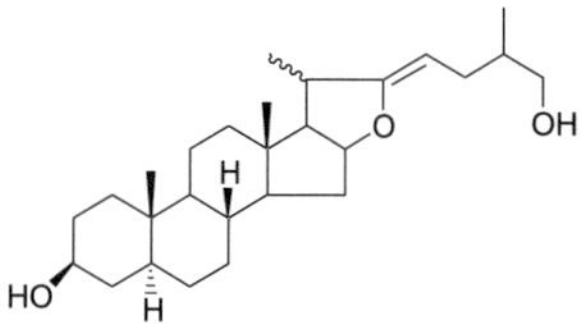

163. 20-OH (20*S*,25*S*)
164. 2α,20-OH (20*S*,25*S*)
165. 2α,20-OH (20*R*,25*S*)
166. 2α-OH,20-OMe (20*R*,25*S*)

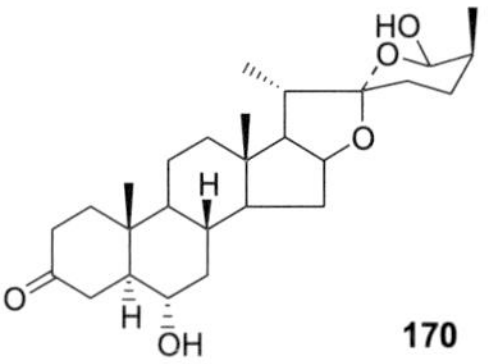

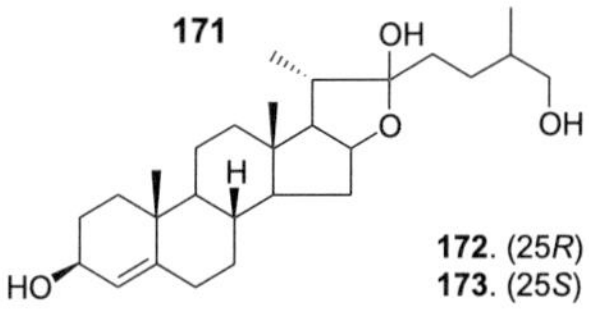

167. 21-OH: Trillenogenin
168. 21-Deoxytrillenogenin

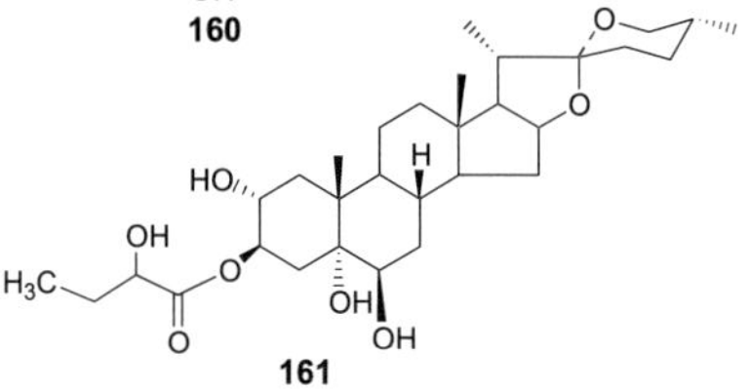

169

170

171

172. (25*R*)
173. (25*S*)

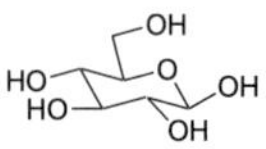

Glucose

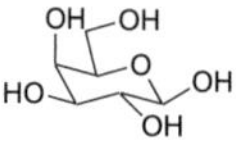

Galactose

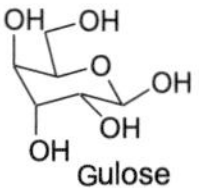

Gulose

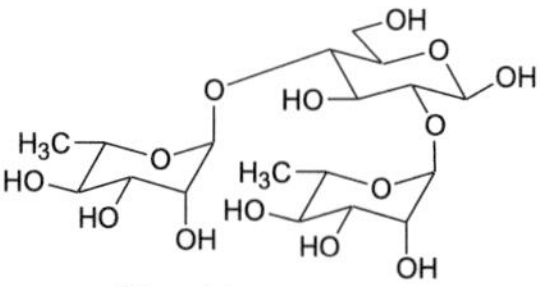

Chacotriose

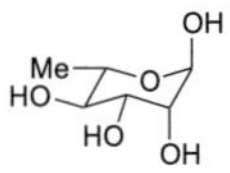

Rhamnose

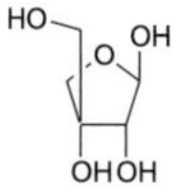

Apiose

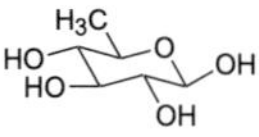

Quinovose

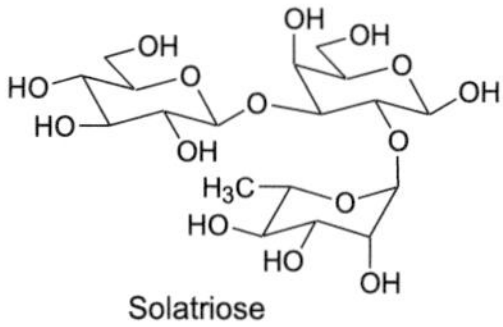

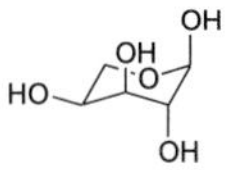

Arabinose

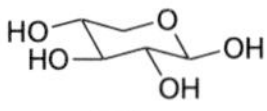

Xylose

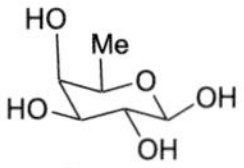

Fucose

Solatriose

7. Conclusion

This review presents recent advances in the techniques used in the isolation and structure elucidation of steroidal saponins as well as a compilation of new steroidal saponins during the last eight years together with their available physical data. About 317 new compounds have been isolated during the period based on 173 genins. Most of these steroidal glycosides possess very complex and highly branched oligosaccharide moieties and present a formidable task for their purification and structure elucidation. HPLC has become an indispensable method of purification of steroidal saponins. Mass spectrometry, particularly ESI-MS, can establish the correct molecular weight and even the oligosaccharide sequence. However, NMR plays the key role in the structural elucidation of the compounds. With the help of 2D-NMR techniques one can now establish complete structures of these saponins without the need for prior acid hydrolysis.

Saponins present in plants or plant products show diverse biological effects in the animal body. Most steroidal saponins exhibit a wide range of cytotoxic effects against cancer cells. The ability to lower the serum cholesterol level has been reported, so also antifungal activity. The effect of steroidal saponins or their derivatives on animal and human reproductive systems is another area that needs attention. These favourable effects and accumulated evidence underline the potential of steroidal glycosides in the development of pharmaceutical preparations. Further developments relating to their use in agriculture need attention too. From the information available in the literature it is difficult to explain the functions of saponins and their structure-activity relationships in biological systems because of the similarity in chemical structures, complexity of physiological reactions involved and the non-availability of pure/homogeneous saponins in sufficient amounts. Alternate methods of investigation may have to be pursued for this. As the study of steroidal saponins has by now provided a reasonable amount of information related to their extraction and structure elucidation, designed compounds may conceivably be prepared in the future though semi-synthesis to allow further biological evaluation of saponins.

Acknowledgement

The authors thank Dr. Basudeb Achari, E. S. (CSIR) for helpful suggestions and encouragement throughout the preparation of the article. Dr. N. P. Sahu is indebted to CSIR for the award of the rank of Emeritus Scientist.

References

1. Williams DH, Stone MJ, Hauck PR, Rahman SK (1989) Why Are Secondary Metabolites (Natural Products) Biosynthesized? J Nat Prod **52**: 1189
2. Hardman R (1987) Recent Developments in our Knowledge of Steroids. Planta Med **53**: 233
3. Hostettmann K, Marston A (1995) Chemistry and Pharmacology of Natural Products: Saponins, p. 1. Cambridge University Press, Cambridge, UK
4. Kemertelidze ÉP, Pkheidze TA (1972) Tigogenin from *Yucca gloriosa*, A Possible Raw Material for the Synthesis of Steroid Hormonal Preparations. Pharm Chem J **6**: 795
5. Mirkin G (1991) Estrogen in Yams. J Amer Med Assoc **265**: 912
6. Djerassi C (1992) Drugs from Third World Plants: The Future. Science **258**: 203
7. Ramberg J, Nugent S (2002) History and Uses of Dioscorea as a Food and Herbal Medicine. Glyco Sci Nutri **3**: 1
8. Marston A, Hostettmann K (1985) Review Article No. 6. Plant Molluscicides. Phytochemistry **24**: 639
9. Mimaki Y, Kuroda M, Fukasawa T, Sashida Y (1999) Steroidal Glycosides from the Bulbs of *Allium jesdianum*. J Nat Prod **62**: 194
10. Miyakoshi M, Tamura Y, Masuda H, Mizutani K, Tanaka O, Ikeda T, Ohtani K, Kasai R, Yamasaki K (2000) Antiyeast Steroidal Saponins from *Yucca schidigera* (Mohave Yucca), A New Anti-Food-Deteriorating Agent. J Nat Prod **63**: 332
11. Mimaki Y, Yokosuka A, Kuroda M, Sashida Y (2001) Cytotoxic Activities and Structure-Cytotoxic Relationships of Steroidal Saponins. Biol Pharm Bull **24**: 1286
12. Křen V, Martinková L (2001) Glycosides in Medicine: The Role of Glycosidic Residue in Biological Activity. Current Med Chem **8**: 1313
13. Francis G, Kerem Z, Makkar HPS, Becker K (2002) The Biological Action of Saponins in Animal Systems: A Review. Br J Nutri **88**: 587
14. Kashibuchi N, Matsubara K, Kitada Y, Suzuki H (1996) Scalp Moisturizer and External Skin Preparation. US Patent No. 5565207
15. Tschesche R, Wulff G (1973) Chemie und Biologie der Saponine. In: Herz W, Grisebach H, Kirby GW (eds.) Fortschr Chem Organ Naturstoffe, Vol. 30, p. 461. Springer, Wien New York
16. Elks J (1971) Steroid Saponins and Sapogenins. In: Coffey S (ed.) Rodd's Chemistry of Carbon Compounds, 2nd Edn, Vol. IIE, p. 1. Elsevier, Amsterdam
17. Elks J (1974) In: Ansell MF (ed.) Rodd's Chemistry of Carbon Compounds (Supplement to the 2nd Edn), Vol. 2D, p. 205. Elsevier, Amsterdam
18. Takeda K (1972) In: Reinhold L, Liwschitz Y (eds.) Progress in Phytochemistry, Vol. 4, p. 287. Interscience, London
19. Mahato SB, Ganguly AN, Sahu NP (1982) Steroid Saponins. Phytochemistry **21**: 959
20. Singh SB, Thakur RS (1983) Recent Advances in the Chemistry of Steroidal Saponins and Their Genins. J Sci Industr Res **42**: 319
21. Voigt G, Hiller K (1987) Advances in the Chemistry and Biology of the Steroid Saponins. Scientia Pharmaceutica **55**: 201
22. Yang Z, Xiao Z (1989) Recent Advances in Chemical Research of Steroidal Saponins (in Chinese). Zhongguo Yaoxue Zazhi **24**: 10
23. Hong J, Jia Z (1995) Recent Progress in Steroidal Saponins (in Chinese). Tianran Chanwu Yanjiu Yu Kaifa **7**: 60
24. Yves S, Baissac Y, Leconte O, Petit P, Ribes G (1996) Steroid Saponins from Fenugreek and Some of Their Biological Properties. Adv Exp Med Biol **405**: 37

25. Mimaki Y, Sashida Y (1996) Steroidal Saponins from the Liliaceae Plants and Their Biological Activities. Adv Exp Med Biol **404**: 101
26. Yang C-R, Li X-C (1996) Bioactive Triterpenoid and Steroid Saponins from Medicinal Plants in Southwest China. Adv Exp Med Biol **404**: 225
27. Kintia PK (1996) Chemistry and Biological Activity of Steroid Saponins from Moldovian Plants. Adv Exp Med Biol **404**: 309
28. Peng JP, Yao XS (1996) 19 New Steroidal Saponins from Allium Plants. Adv Exp Med Biol **404**: 511
29. Zhang JB, Yu B, Hui YZ (2000) Recent Progress in Research of Furostanol Saponins. Youji Huaxue **20**: 663
30. Sun Q, Yong J, Zhao Y (2002) Steroid Saponins with Biological Activities. Zhongcaoyao **33**: 276
31. Agrawal PK (1996) A Systematic Approach for the Determination of the Molecular Structure of Steroid Saponins. Adv Exp Med Biol **405**: 299
32. Agrawal PK, Jain DC, Pathak AK (1995) NMR Spectral Investigation. Part 37. NMR Spectroscopy of Steroidal Sapogenins and Steroidal Saponins: An Update. Mag Reson Chem **33**: 923
33. Mahato SB, Sahu NP, Pal BC, Chakravarti RN (1977) Constitution of New Steroidal Saponins Isolated from *Kallstroemia pubescens*. Indian J Chem **15B**: 445
34. Mahato SB, Sahu NP, Pal BC (1978) New Steroidal Saponins from *Dioscorea floribunda*: Structures of Floribundasaponins C, D, E and F. Indian J Chem **16B**: 350
35. Mandal D, Banerjee S, Mondal NB, Chakravarty AK, Sahu NP (2006) Steroidal Saponins from the Fruits of *Asparagus racemosus*. Phytochemistry **67**: 1316
36. Ohtsuki T, Koyano T, Kowithayakorn T, Sakai S, Kawahara N, Goda Y, Yamaguchi N, Ishibashi M (2004) New Chlorogenin Hexasaccharide Isolated from *Agave fourcroydes* with Cytotoxic and Cell Inhibitory Activities. Bioorg Med Chem **12**: 3841
37. Pettit GR, Zhang Q, Pinilla V, Hoffmann H, Knight JC, Doubek DL, Chapuis J-C, Pettit RK, Schmidt JM (2005) Antineoplastic Agents. 534. Isolation and Structure of Sansevistatins 1 and 2 from the African *Sansevieria ehrenbergii*. J Nat Prod **68**: 729
38. Sautour M, Miyamoto T, Lacaille-Dubois M-A (2005) Steroidal Saponins from *Smilax medica* and Their Antifungal Activity. J Nat Prod **68**: 1489
39. Yoshimitsu H, Nishida M, Nohara T (2003) Steroidal Glycosides from the Fruits of *Solanum abutiloides*. Phytochemistry **64**: 1361
40. Zhang J, Ma B, Kang L, Yu H, Yang Y, Yan X, Dong F (2006) Furostanol Saponins from the Fresh Rhizomes of *Polygonatum kingianum*. Chem Pharm Bull **54**: 931
41. Marker RE, Lopez J (1947) Steroidal Sapogenins, No. 165. Structure of the Sapogenin Glycosides. J Amer Chem Soc **69**: 2389
42. Fieser L, Fieser M (1959) Steroids, p. 832. Reinhold, New York
43. Nohara T, Miyahara K, Kawasaki T (1975) Steroid Saponins and Sapogenins of Underground Part of *Trillium kamtschaticum* Pall. II. Pennogenin and Kryptogenin 3-*O*-Glycosides and Related Compounds. Chem Pharm Bull **23**: 872
44. Stahl E (1962) Dünnschicht-Chromatographie, pp. 498, 503. Springer, Berlin
45. Marker RE, Turner DL (1940) The Oxidation of Pregnane Triols. J Amer Chem Soc **62**: 2540
46. Nohara T, Miyahara K, Komori T, Kawasaki T (1975) Structure of Novel-type Steroid Glycoside. Tetrahedron Lett **49**: 4381
47. Nohara T, Komori T, Kawasaki T (1980) Steroid Saponins and Sapogenins of Underground Parts of *Trillium kamtschaticum* Pall. III. On the Structure of Novel

Type of Steroid Glycoside, Trillenoside A, An 18-Norspirostanol Oligoside. Chem Pharm Bull **28**: 1437
48. Fukuda N, Imamura N, Saito E, Nohara T, Kawasaki T (1981) Steroid Saponins and Sapogenins of Underground Parts of *Trillium kamtschaticum* Pall. IV. Additional Oligoglycosides of 18-Norspirostane Derivatives and Other Steroidal Constituents. Chem Pharm Bull **29**: 325
49. Ono M, Yanai Y, Ikeda T, Okawa M, Nohara T (2003) Steroids from Underground Parts of *Trillium kamtschaticum*. Chem Pharm Bull **51**: 1325
50. Nakano K, Nohara T, Tomimatsu T, Kawasaki T (1982) A Novel 18-Norspirostanol Bisdesmoside from *Trillium tschonoskii*. J Chem Soc Chem Comm: 789
51. Nakano K, Nohara T, Tomimatsu T, Kawasaki T (1983) 18-Norspirostanol Derivatives from *Trillium tschonoskii*. Phytochemistry **22**: 1047
52. Ono M, Hamada T, Nohara T (1986) An 18-Norspirostanol Glycoside from *Trillium tschonoskii*. Phytochemistry **25**: 544
53. Nohara T, Ito Y, Seike H, Komori T, Moriyama M, Gomita Y, Kawasaki T (1982) Study of the Constituents of *Paris quadrifolia* L. Chem Pharm Bull **30**: 1851
54. Yokosuka A, Mimaki Y, Sashida Y (2002) Four New 3,5-Cyclosteroidal Saponins from *Dracaena surculosa*. Chem Pharm Bull **50**: 992
55. Becker RC, Bianchi E, Cole JR (1972) A Phytochemical Investigation of *Yucca schottii* (Liliaceae). J Pharm Sci **61**: 1665
56. Amoros M, Girre RL (1987) Structure of Two Antiviral Triterpene Saponins from *Anagallis arvensis*. Phytochemistry **26**: 787
57. Meselhy MR, Aboutabl EA (1997) Hopane-Type Saponins from *Polycarpon succulentum* Growing in Egypt. Phytochemistry **44**: 925
58. Meselhy MR (1998) Hopane-Type Saponins from *Polycarpon succulentum*. Phytochemistry **48**: 1415
59. Krider MM, Branaman JR, Wall ME (1955) Steroidal sapogenins XVIII. Partial Hydrolysis of Steroidal Saponins of *Yucca schidigera*. J Amer Chem Soc **77**: 1238
60. Krokhmalyuk VV, Kintya PK (1977) Steroid Saponins X. Glycosides of *Allium narcissiflorum*: The Structure of Glycosides A and B. Chem Nat Comp **12**: 46
61. Mahato SB, Sahu NP, Ganguly AN (1981) Steroidal Saponins from *Dioscorea floribunda*: Structures of Floribundasaponins A and B. Phytochemistry **20**: 1943
62. Mimaki Y, Kanmoto T, Sashida Y, Nishino A, Satomi Y, Nishino H (1996) Steroidal Saponins from the Underground Parts of *Chlorophytum comosum* and Their Inhibitory Activity on Tumor Promoter-Induced Phospholipid Metabolism of Hela Cells. Phytochemistry **41**: 1405
63. Mimaki Y, Kuroda M, Kameyama A, Yokosuka A, Sashida Y (1998) Steroidal Saponins from the Underground Parts of *Ruscus aculeatus* and Their Cytostatic Activity on HL-60 Cells. Phytochemistry **48**: 485
64. Sahu NP, Koike K, Banerjee S, Achari B, Nikaido T (2001) Triterpenoid Saponins from *Mollugo spergula*. Phytochemistry **58**: 1177
65. Mimaki Y, Kuroda M, Kameyama A, Yokosuka A, Sashida Y (1998) New Steroidal Constituents of the Underground Parts of *Ruscus aculeatus* and Their Cytostatic Activity on HL-60 Cells. Chem Pharm Bull **46**: 298
66. Pocsi I, Kiss L, Hughes MA, Nanasi P (1989) Kinetic Investigation of the Substrate Specificity of the Cyanogenic-β-glucosidase (Linamarase) of White Clover. Arch Biochem Biophys **272**: 496
67. Poulton JE (1990) Cyanogenesis in Plants. Plant Physiol **94**: 401
68. Sue M, Ishihara A, Iwamura H (2000) Purification and Characterization of a β-Glucosidase from Rye (*Secale cereale* L.) Seedlings. Plant Science **155**: 67

69. Hrmova M, Harvey AJ, Wang J, Shirley NJ, Jones GP, Stone BA, Hoj PB, Fincher GB (1996) Barley β-D-Glucan Exohydrolase with β-D-Glucosidase Activity. Purification, Characterization, and Determination of Primary Structure from a cDNA Clone. J Biol Chem **271**: 5277
70. Akiyama T, Kaku H, Shibuya NA (1998) Cell Wall Bound β-Glucosidase from Germinated Rice: Purification and Properties. Phytochemistry **48**: 49
71. Svasti J, Srisomsap C, Techasakul S, Surarit R (1999) Dalchochinin-8'-O-β-D-Glucoside and Its β-Glucosidase Enzyme from *Dalbergia cochinchinensis*. Phytochemistry **50**: 739
72. Nisius A (1988) The Stromacentre in Avena Plastids: An Aggregation of β-Glucosidase Responsible for the Activation of Oat-Leaf Saponins. Planta **173**: 474
73. Gus-Mayer S, Brunner H, Schneider-Poetsch HA, Rüdiger W (1994) Avenacosidase from Oat: Purification, Sequence Analysis and Biochemical Characterization of New Member of the BGA Family of β-Glucosidases. Plant Mol Biol **26**: 909
74. Inoue K, Ebizuka Y (1996) Purification and Characterization of Furostanol Glycoside 26-O-Glucosidase from *Costus speciosus* Rhizomes. FEBS Lett **378**: 157
75. Arthan D, Kittakoop P, Esen A, Svasti J (2006) Furostanol Glycoside 26-O-Glucosidase from the Leaves of *Solanum torvum*. Phytochemistry **67**: 27
76. Arthan D, Svasti J, Kittakoop P, Pittayakhachonwut D, Tanticharoen M, Thebtaranonth Y (2002) Antiviral Isoflavonoid Sulfate and Steroidal Glycosides from the Fruits of *Solanum torvum*. Phytochemistry **59**: 459
77. Vollerner YS, Abdullaev ND, Gorovits MB, Abubakirov NK (1984) Steroid Saponins and Sapogenins of *Allium*. XIX. The Structure of Karatavigenin C. Chem Nat Comp **19**: 699
78. Kintya PK, Stamova AI, Bakinovskii LB, Krokhmalyuk VV (1978) Steroid Glycosides (XXI). The Structure of Polygonatoside E' and Protopolygonatoside E' from the Leaves of *Polygonatum latifolium*. Chem Nat Comp **14**: 290
79. Li XC, Yang CR, Matsuura H, Kasai R, Yamasaki K (1993) Steroidal Glycosides from *Polygonatum prattii*. Phytochemistry **33**: 465
80. Son KH, Do JC (1990) Steroidal Saponins from the Rhizomes of *Polygonatum sibiricum*. J Nat Prod **53**: 333
81. Mimaki Y, Kuroda M, Fukasawa T, Sashida Y (1999) Steroidal Saponins from the Bulbs of *Allium karataviense*. Chem Pharm Bull **47**: 738
82. Mackie AM, Turner AB (1970) Partial Characterization of a Biologically Active Steroid Glycoside from the Starfish *Marthasterias glacialis*. Biochem J **117**: 543
83. Smith F, Unrau AM (1959) On the Presence of 1 → 6 Linkages in Laminarin. Chem Ind 881
84. Goldstein IJ, Hay GW, Lewis BA, Smith F (1965). In: Whistler HL (ed.) Methods in Carbohydrate Chemistry, Vol. 5, p. 361. Academic Press, New York
85. Plock A, Beyer G, Hiller K, Gründemann E, Krause E, Nimtz M, Wray V (2001) Application of MS and NMR to the Structure Elucidation of Complex Sugar Moieties of Natural Products: Exemplified by the Steroidal Saponin from *Yucca filamentosa* L. Phytochemistry **57**: 489
86. Hayashi K, Iida I, Nakao Y, Kaneko Y (1988) Four Pregnane Glycosides, Boucerosides AI, AII, BI and BIII from *Boucerosia aucheriana*. Phytochemistry **27**: 3919
87. Tsukamoto S, Hayashi K, Kaneko K, Mitsuhashi H (1986) Studies on the Constituents of Asclepiadaceae Plants. LXV. The Optical Resolution of D- and L-Cymaroses. Chem Pharm Bull **34**: 3130

88. König WA, Benecke I, Bretting H (1981) Gas Chromatographic Separation of Carbohydrate Enantiomers on a New Chiral Stationary Phase. Angew Chem Int Ed Engl **20**: 693
89. Chang M, Meyers HV, Nakanishi K, Ojika M, Park JH, Park MH, Takeda R, Vazquez JT, Wiesler WT (1989) Microscale Structure Determination of Oligosaccharides by the Exciton Chirality Method. Pure Appl Chem **61**: 1193
90. Klyne W (1950) The Configuration of the Anomeric Carbon Atoms in Some Cardiac Glycosides. Biochem J **47**: xli
91. Mahato SB, Sahu NP, Ganguly AN, Miyahara K, Kawasaki T, Tanaka O (1981) Steroidal Glycosides of *Tribulus terrestris* Linn. J Chem Soc Perkin Trans I: 2405
92. Eggert H, Djerassi C (1975) ^{13}C NMR Spectra of Sapogenins. Tetrahedron Lett **16**: 3635
93. Mahato SB, Sahu NP, Ganguly AN, Kasai YR, Tanaka O (1980) Steroidal Alkaloids from *Solanum khasianum*: Application of ^{13}C NMR Spectroscopy to Their Structural Elucidation. Phytochemistry **19**: 2017
94. Seo S, Tomita Y, Tori K, Yoshimura Y (1978) Determination of the Absolute Configuration of a Secondary Hydroxy Group in a Chiral Secondary Alcohol Using Glycosidation Shifts in Carbon-13 Nuclear Magnetic Resonance Spectroscopy. J Amer Chem Soc **100**: 3331
95. Stothers JB (1972) Carbon-13 NMR Spectroscopy. Academic Press, New York
96. Kasai R, Okihara M, Asakawa J, Mizutani K, Tanaka O (1979) ^{13}C NMR Study of α- and β-Anomeric Pairs of D-Mannopyranosides and L-Rhamnopyranosides. Tetrahedron **35**: 1427
97. Kasai R, Suzuo M, Asakawa J, Tanaka O (1977) Carbon-13 Chemical Shifts of Isoprenoid-β-D-glucopyranosides and β-D-Mannopyranosides. Stereochemical Influences of Aglycone Alcohols. Tetrahedron Lett **18**: 175
98. Tori K, Seo S, Oshimura Y, Arita H, Tomita Y (1977) Glycosidation Shifts in Carbon-13 NMR Spectroscopy: Carbon-13 Signal Shifts from Aglycone and Glucose to Glucoside. Tetrahedron Lett **18**: 179
99. Massiot G, Lavaud C, Guillaume D, Le Men-Olivier L, Van-Binst G (1986) Identification and Sequencing of Sugars in Saponins Using 2D ^{1}HNMR Spectroscopy. J Chem Soc Chem Comm: 1485
100. Wolfender J-L, Rodriguez S, Hostettmann K (1998) Liquid Chromatography Coupled to Mass Spectrometry and Nuclear Magnetic Resonance Spectroscopy for the Screening of Plant Constituents. J Chromatography A **794**: 299
101. Williams DH, Bradley G, Bojesen G, Santokaran S, Taylor LCE (1981) Fast Atom Bombardment Mass Spectrometry: A Powerful Technique for the Study of Polar Molecules. J Amer Chem Soc **103**: 5700
102. Fenselau C (1984) Fast Atom Bombardment and Middle Molecule Mass Spectrometry. J Nat Prod **47**: 215
103. Zhou ZL, Aquino R, De Simone F, Dini A, Schettino O, Pizza C (1988) Oligofurostanosides from *Asparagus cochinchinensis*. Planta Med **54**: 344
104. Inoue T, Mikaki Y, Sashida Y, Nishino A, Satomi Y, Nishino H (1995) Steroidal Glycosides from *Allium macleanii* and *A. senescens*, and Their Inhibitory Activity on Tumour Promoter-Induced Phospholipid Metabolism of Hela Cells. Phytochemistry **40**: 521
105. Yan W, Ohtani K, Kasai R, Yamasaki K (1996) Steroidal Saponins from Fruits of *Tribulus terrestris*. Phytochemistry **42**: 1417
106. Debella A, Haslinger E, Kunert O, Michi C, Abebe D (1999) Steroidal Saponins from *Asparagus africanus*. Phytochemistry **51**: 1069

107. Lattimer RP, Schulten HR (1989) Field Ionization and Field Desorption Mass Spectrometry: Past, Present and Future. Anal Chem **61**: 1201A
108. Komori T, Kawasaki T, Schulten HR (2005) Field Desorption and Fast Atom Bombardment Mass Spectrometry of Biologically Active Natural Oligoglycosides. Mass Spectro Reviews **4**: 255
109. Sundqvist B, Roepstorff P, Fohlman J, Hedin A, Hakansson P, Kamensky M, Lindberg M, Salehpoour M, Save G (1984) Molecular Weight Determinations of Proteins by Californium Plasma Desorption Mass Spectrometry. Science **226**: 696
110. Pilipenko VV, Sukhodub LF, Aksyonov SA, Kalinkevich AN, Kintia PK (2000) ^{252}Cf Plasma Desorption Mass Spectrometric Study of Interactions of Steroid Glycosides with Amino Acids. Rapid Commun Mass Spectrom **14**: 819
111. Karas M, Hillenkamp F (1988) Laser Desorption Ionization of Proteins with Molecular Masses Exceeding 10,000 Daltons. Anal Chem **60**: 2299
112. Liu SY, Cui M, Liu ZQ, Song FR (2004) Structural Analysis of Saponins from Medicinal Herbs Using Electrospray Ionization Tandem Mass Spectrometry. J Amer Soc Mass Spectrom **15**: 133
113. McLafferty FW, Fridriksson EK, Horn DM, Lewis MA, Zubarev RA (1999) Techview: Biochemistry, Biomolecules Mass Spectrometry. Science **284**: 1289
114. Wilm M, Shevchenko A, Houthaeve T, Breit S, Schweigerer L, Fotsis T, Mann M (1996) Femtomole Sequencing of Proteins from Polyacrylamide Gels by Nano-Electrospray Mass Spectrometry. Nature **379**: 466
115. Shen X, Perreault H (1999) Electrospray Ionization Mass Spectrometry of 1-Phenyl-3-methyl-5-pyrazolone Derivatives of Neutral and N-Acetylated Oligosaccharides. J Mass Spectrom **34**: 502
116. Chai W, Piskarev V, Lawson AM (2001) Negative-Ion Electrospray Mass Spectrometry of Neutral Underivatized Oligosaccharides. Anal Chem **73**: 651
117. Putalun W, Tanaka H, Muranaka T, Shoyama Y (2002) Determination of Aculeatisides Based on Immunoassay Using a Polyclonal Antibody Against Aculeatiside A. Analyst **127**: 1328
118. Fang SP, Hao CY, Sun WX, Liu ZQ, Liu SY (1998) Rapid Analysis of Steroidal Saponin Mixture Using Electrospray Ionization Mass Spectrometry Combined with Sequential Tandem Mass Spectrometry. Rapid Commun Mass Spectrom **12**: 589
119. Fang SP, Hao CY, Liu ZQ, Song FR, Liu SY (1999) Application of Electrospray Ionization Mass Spectrometry Combined with Sequential Tandem Mass Spectrometry Techniques for the Profiling of Steroidal Saponin Mixture Extracted from *Tribulus terrestris*. Planta Med **65**: 68
120. Cui M, Sun WX, Song FR, Liu ZQ, Liu SY (1999) Multi-Stage Mass Spectrometric Studies of Triterpenoid Saponins in Crude Extracts from *Acanthopanax senticosus* Harms. Rapid Commun Mass Spectrom **13**: 873
121. Van Setten DC, Zomer G, Van DeWerken G, Wiertz EJHJ, Leeflang BR, Kamerling JP (2000) Ion Trap Multiple-Stage Tandem Mass Spectrometry as a Pre-NMR Tool in the Structure Elucidation of Saponins. Phytochem Anal **11**: 190
122. Guo MQ, Song FR, Bai Y, Liu ZQ, Liu SY (2002) Rapid Analysis of a Triterpenoid Saponin Mixture from Plant Extracts by Electrospray Ionization Multi-Stage Tandem Mass Spectrometry (ESI-MS). Anal Sci **18**: 481
123. Song FR, Cui M, Liu ZQ, Yu B, Liu SY (2004) Multiple-Stage Tandem Mass Spectrometry for Differentiation of Isomeric Saponins. Rapid Commun Mass Spectrom **18**: 2241

124. Brobera S, Nord LI, Kenne L (2004) Oligosaccharide Sequences in Quillaja Saponins by Electrospray Ionization Ion Trap Multi-Stage Mass Spectrometry. J Mass Spectrom **39**: 691
125. Li R, Zhou Y, Wu Z, Ding L (2006) ESI-Q TOF-MS/MS and APCI-IT-MS/MS Analysis of Steroid Saponins from the Rhizomes of *Dioscorea panthaica*. J Mass Spectrom **41**: 1
126. Liang F, Li L-J, Abliz Z, Yang Y-C, Shi J-G (2002) Structural Characterization of Steroidal Saponins by Electrospray Ionization and Fast-Atom Bombardment Tandem Mass Spectrometry. Rapid Commun Mass Spectrom **16**: 1168
127. Agrawal PK (2003) Spectral Assignments and Reference Data: 25R/25S Stereochemistry of Spirostane-Type Steroidal Sapogenins and Steroidal Saponins via Chemical Shift of Geminal Protons of Ring-F. Magn Reson Chem **41**: 965
128. Agrawal PK (2005) Assigning Stereodiversity of the 27-Me Group of Furostane-Type Steroidal Saponins via NMR Chemical Shifts. Steroids **70**: 715
129. Ohtsuki T, Sato M, Koyano T, Kowithayakorn T, Kawahara N, Goda Y, Ishibashi M (2006) Steroidal Saponins from *Calamus insignis*, and Their Cell Growth and Cell Cycle Inhibitory Activities. Bioorg Med Chem **14**: 659
130. Haraguchi M, Mimaki Y, Motidome M, Morita H, Takeya K, Itokawa H, Yokosuka A, Sashida Y (2000) Steroidal Saponins from the Leaves of *Cestrum sendtenerianum*. Phytochemistry **55**: 715
131. Sang S, Mao S, Lao A, Chan Z, Ho C-T (2001) Four New Steroidal Saponins from the Seeds of *Allium tuberosum*. J Agric Food Chem **49**: 1475
132. Agrawal PK (1992) NMR Spectroscopy in the Structural Elucidation of Oligosaccharides and Glycosides. Phytochemistry **31**: 3307
133. Sahu NP, Koike K, Jia Z, Nikaido T (1995) Novel Triterpenoid Saponins from *Mimusops elengi*. Tetrahedron **51**: 13435
134. Doddwell DM, Pegg DT, Bendall MR (1982) Distortionless Enhancement of NMR Signals by Polarization Transfer. J Magn Reson **48**: 323
135. Jin J-M, Zhang Y-J, Yang C-R (2004) Spirostanol and Furostanol Glycosides from the Fresh Tubers of *Polianthes tuberosa*. J Nat Prod **67**: 5
136. Bedir E, Khan IA (2000) New Steroidal Glycosides from the Fruits of *Tribulus terrestris*. J Nat Prod **63**: 1699
137. Yokosuka A, Mimaki Y, Sashida Y (2002) Steroidal and Pregnane Glycosides from the Rhizomes of *Tacca chantrieri*. J Nat Prod **65**: 1293
138. Agrawal PK, Bunsawansong P, Morris GA (1997) Complete Assignment of the ^{1}H and ^{13}C NMR Spectra of Steroidal Sapogenins: Smilagenin and Sarsasapogenin. Magn Reson Chem **35**: 441
139. Braunschweiler L, Ernst RR (1983) Coherence Transfer by Isotropic Mixing: Application to Proton Correlation Spectroscopy. J Magn Reson **53**: 521
140. Kessler H, Gehrke M, Griesinger C (1988) Two-Dimensional NMR Spectroscopy: Background and Overview of the Experiments. Angew Chem Int Ed Engl **27**: 490
141. Davis DG, Bax A (1985) Assignment of Complex ^{1}H NMR Spectra via Two-Dimensional Homonuclear Hartmann-Hahn Spectroscopy. J Amer Chem Soc **107**: 2820
142. Marx RS, Glaser J (2003) Spins Swing Like Pendulums Do: An Exact Classical Model for TOCSY Transfer in Systems of Three Isotropically Coupled Spins 1/2. J Magn Reson **164**: 338
143. Bax A, Morris GA (1981) An Improved Method for Heteronuclear Chemical Shift Correlation by Two-Dimensional NMR. J Magn Reson **42**: 501

144. Summers MF, Marzilli LG, Bax A (1986) Complete Proton and Carbon-13 Assignments of Coenzyme B12 Through the Use of New Two-Dimensional NMR Experiments. J Amer Chem Soc **108**: 4285
145. Bax A, Summers MF (1986) Proton and Carbon-13 Assignments from Sensitivity-Enhanced Detection of Heteronuclear Multiple-Bond Connectivity by 2D Multiple Quantum NMR. J Amer Chem Soc **108**: 2093
146. Macura S, Ernst RR (1980) Elucidation of Cross Relaxation in Liquids by Two-Dimensional NMR Spectroscopy. Mol Phys **41**: 95
147. Macura S, Huang Y, Suter D, Ernst RR (1981) Two-Dimensional Chemical Exchange and Cross-Relaxation Spectroscopy of Coupled Nuclear Spins. J Magn Reson **43**: 259
148. Claridge TDW (1999) Correlations Through Space: The Nuclear Overhauser Effect. In: Baldwin JE, Williams RM (eds.) High-Resolution NMR Techniques in Organic Chemistry, Vol. 19, p. 277. Elsevier Science, Oxford, UK
149. D'Auria MV, Giannini C, Zampella A, Minale L, Deditus C, Roussakis C (1998) Crellastatin A: A Cytotoxic Bis-Steroid Sulfate from the Vanuatu Marine Sponge *Crella* sp. J Org Chem **63**: 7382
150. Croasmun WR, Carlson RMK (1994) Steroidal Structural Analysis by Two-Dimensional NMR. In: Croasmun WR, Carlson RMK (eds.) Two-Dimensional NMR Spectroscopy Applications for Chemists and Biochemists, 2nd Edn, p. 785. Wiley-VCH, Weinheim
151. Bross-Walch N, Kühn T, Moskau D, Zerbe O (2005) Strategies and Tools for Structure Determination of Natural Products Using Modern Methods of NMR Spectroscopy. Chem Biodivers **2**: 147
152. Rance M, Sørensen OW, Bodenhausen G, Wagner G, Ernst RR, Wüthrich K (1983) Improved Spectral Resolution in COSY NMR Spectra of Proteins via Double Quantum Filtering. Biochem Biophys Res Commun **117**: 479
153. Vuister GW, DeWarrd P, Boelens R (1989) The Use of 3D NMR in Structural Studies of Oligosaccharides. J Amer Chem Soc **111**: 772
154. Bock K, Pedersen C, Pedersen H (1984) Carbon-13 NMR Data for Oligosaccharides. Adv Carbohydr Chem Biochem **42**: 193
155. Sahu NP, Achari B (2001) Advances in Structural Determination of Saponins and Terpenoid Glycosides. Curr Org Chem **5**: 315
156. Bothner-By AA, Stephens RL, Lee J, Warren CD, Jeanloz RW (1984) Structure Determination of a Tetrasaccharide: Transient Nuclear Overhauser Effects in the Rotating Frame. J Amer Chem Soc **106**: 811
157. Bax A, Davis DG (1985) Practical Aspects of Two-Dimensional Transverse NOE Spectroscopy. J Magn Reson **63**: 207
158. Jia Z, Koike K, Nikaido T (1999) Saponarioside C, the First α-D-Galactose Containing Triterpenoid Saponin, and Five Related Compounds from *Saponaria officinalis*. J Nat Prod **62**: 449
159. George AJ (1965) Legal Status and Toxicity of Saponins. Food Cosmet Toxicol **3**: 85
160. El Izzi A, Benie T, Thieulant M-L, Le Men-Olivier L, Duval J (1992) Stimulation of LH Release from Cultured Pituitary Cells by Saponins of *Petersianthus macrocarpus*: A Permeabilising Effect. Planta Med **58**: 229
161. Authi KS, Rao GHR, Evenden BJ, Crawford N (1988) Action of Guanosine 5'-(beta-thio)Diphosphate on Thrombin-Induced Activation and Calcium Mobilization in Saponin-Permeabilized and Intact Human Platelets. Biochem J **255**: 885
162. Plock A, Sokolowska W, Presber W (2001) Application of Flow Cytometry and Microscopical Methods to Characterize the Effect of Herbal Drugs on *Leishmania* spp. Exp Parasitol **97**: 1451

163. Kensil CR (1996) Saponins as Vaccine Adjuvants. Crit Rev Ther Drug **13**: 1
164. Barr IG, Sjolander A, Cox JC (1998) ISCOMs and Other Saponin Based Adjuvants. Adv Drug Deli Reviews **32**: 247
165. Sen S, Makkar HPS, Becker K (1998) Alfalfa Saponins and Their Implication in Animal Nutrition. J Agric Food Chem **46**: 131
166. Yoshiki Y, Kudou S, Okubo K (1998) Relationship Between Chemical Structures and Biological Activities of Triterpenoid Saponins from Soybean (Review). Biosci Biotechnol Biochem **62**: 2291
167. Křen V, Martinková L (2001) Glycosides in Medicine: The Role of Glycosidic Residue in Biological Activity. Curr Med Chem **8**: 1313
168. Francis G, Kerem Z, Makkar HPS, Becker K (2002) The Biological Action of Saponins in Animal Systems: A Review. Brit J Nutr **88**: 587
169. Mimaki Y, Kuroda M, Kameyama A, Yokosuka A, Sashida Y (1998) Steroidal Saponins from the Rhizomes of *Hosta sieboldii* and Their Cytostatic Activity on HL-60 Cells. Phytochemistry **48**: 1361
170. Sargent JM, Taylor CG (1989) Appraisal of the MTT Assay as a Rapid Test of Chemo-Sensitivity in Acute Myeloid Leukaemia. Brit J Cancer **60**: 206
171. Mimaki Y, Kuroda M, Fukasawa T, Sashida Y (1999) Steroidal Glycosides from the Bulbs of *Allium jesdianum*. J Nat Prod **62**: 194
172. Monks A, Scudiero D, Skehan P, Shoemaker R, Paull K, Vistica D, Hose C, Langley J, Cronise P, Vaigro-Wolff A, Gray-Goodrich M, Campbell H, Mayo J, Boyd M (1991) Feasibility of a High-Flux Anticancer Drug Screen Using a Diverse Panel of Cultured Human Tumor Cell Lines. J Natl Cancer Inst **83**: 757
173. Mimaki Y, Yokosuka A, Sashida Y (2000) Steroidal Glycosides from the Aerial Parts of *Polianthes tuberosa*. J Nat Prod **63**: 1519
174. Mimaki Y, Kuroda M, Ide A, Kameyama A, Yokosuka A, Sashida Y (1999) Steroidal Saponins from the Aerial Parts of *Dracaena draco* and Their Cytostatic Activity on HL-60 cells. Phytochemistry **50**: 805
175. Nohara T, Miyahara K, Kawasaki T (1975) Steroid Saponins and Sapogenins of Underground Parts of *Trillium kamtschaticum* Pall. II. Pennogenin- and Kryptogenin 3-*O*-Glycosides and Related Compounds. Chem Pharm Bull **23**: 872
176. Hu K, Yao X (2001) Methyl Protogracillin (NSC-698792): The Spectrum of Cytotoxicity Against 60 Human Cancer Cell Lines in the National Cancer Institute's Anticancer Drug Screen Panel. Anticancer Drugs **12**: 541
177. Hu K, Yao X (2003) The Cytotoxicity of Methyl Protoneogracillin (NSC-698793) and Gracillin (NSC-698787), Two Steroidal Saponins from the Rhizomes of *Dioscorea collettii* var. *hypoglauca*, Against Human Cancer Cells *in vitro*. Phytother Res **17**: 620
178. Paull KD, Shoemaker RH, Hodes L, Monks A, Scudiero DA, Rubinstein L, Plowman J, Boyd MR (1989) Display and Analysis of Patterns of Differential Activity of Drugs Against Human Tumor Cell Lines: Development of Mean Graph and COMPARE Algorithm. J Natl Cancer Inst **81**: 1088
179. Weinstein JN, Myers TG, O'Connor PM, Friend SH, Fornace AJ Jr, Khon KW, Fojo T, Bates SE, Rubinstein LV, Anderson NL, Buolamwini JK, Van Osdol WW, Monks AP, Scudiero DA, Sausville EA, Zaharevitz DW, Bunow B, Viswanadhan VN, Johnson GS, Wittes RE, Paull KD (1997) An Information-Intensive Approach to the Molecular Pharmacology of Cancer. Science **275**: 343
180. Mimaki Y, Watanabe K, Ando Y, Sakuma C, Sashida Y, Furuya S, Sakagami H (2001) Flavonol Glycosides and Steroidal Saponins from the Leaves of *Cestrum nocturnum* and Their Cytotoxicity. J Nat Prod **64**: 17

181. Sata N, Matsunaga S, Fusetani N, Nishikawa H, Takamura S, Saito T (1998) New Antifungal and Cytotoxic Steroidal Saponins from the Bulbs of an Elephant Garlic Mutant. Biosci Biotechnol Biochem **62**: 1904
182. Yokosuka A, Mimaki Y, Sashida Y (2002) Spirostanol Saponins from the Rhizomes of *Tacca chantrieri* and Their Cytotoxic Activity. Phytochemistry **61**: 73
183. Mimaki Y, Watanabe K, Sakagami H, Sashida Y (2002) Steroidal Glycosides from the Leaves of *Cestrum nocturnum*. J Nat Prod **65**: 1863
184. Ahn K-J, Kim CY, Yoon K-D, Ryu MY, Cheong JH, Chin Y-W, Kim J (2000) Steroidal Saponins from the Rhizomes of *Polygonatum sibiricum*. J Nat Prod **69**: 360
185. Kim G-S, Kim H-T, Seong J-D, Oh S-R, Lee C-O, Bang J-K, Seong N-S, Song K-S (2005) Cytotoxic Steroidal Saponins from the Rhizomes of *Asparagus oligoclonos*. J Nat Prod **68**: 766
186. Carmichael J, Degraff WG, Gazdar AF, Minna JD, Mitchell JB (1987) Evaluation of Tetrazolium-Based Semiautomated Colorimetric Assay: Assessment of Chemosensitivity Testing. Cancer Res **47**: 936
187. Boyd MR (1997) Drug Development: Preclinical Screening, Clinical Trial and Approval. In Teicher B (ed.) Cancer Drug Discovery and Development, Vol. 2, p 23. Humana Press, Totowa, NJ
188. Zhou X, He X, Wang G, Gao H, Zhou G, Ye W, Yao X (2006) Steroidal Saponins from *Solanum nigrum*. J Nat Prod **69**: 1158
189. Ikeda T, Tsumagari H, Honbu T, Nohara T (2003) Cytotoxic Activity of Steroidal Glycosides from *Solanum* Plants. Biol Pharm Bull **26**: 1198
190. Kinjo M, Oka K, Naito S, Kohga S, Tanaka K, Oboshi S, Hayata Y, Yasumoto K (1979) Thromboplastic and Fibrinolytic Activities of Cultured Human Cancer Cell Lines. Brit J Cancer **39**: 15
191. Brattain MG, Fine WD, Khaled FM, Thompson J, Brattain DE (1981) Heterogeneity of Malignant Cells from a Human Colonic Carcinoma. Cancer Res **41**: 1751
192. Hernández JC, León F, Quintana J, Estévez F, Bermejo J (2004) Icogenin, a New Cytotoxic Steroidal Saponin Isolated from *Dracaena draco*. Bioorg Med Chem **12**: 4423
193. Mosmann T (1983) Rapid Colorimetric Assay for Cellular Growth and Survival: Application to Proliferation and Cytotoxicity Assays. J Immunol Methods **65**: 55
194. Tewari M, Quan LT, O'Rourke K, Desnoyers S, David ZE, Guy RR, Poirier G, Salvesen GS, Dixit VM (1995) Yama/CPP. 32 β, A Mammalian Homolog of CED-3, is a CrmA Inhibitable Protease that Cleaves the Death Substrate Poly (ADP-ribose) Polymerase. Cell **81**: 801
195. Germain M, Affar EB, D'Amours D, Dixit VM, Salvesen GS, Poirier GG (1999) Cleavage of Automodified Poly(ADP-ribose) Polymerase During Apoptosis. J Biol Chem **274**: 28379
196. Tran QL, Tezuka Y, Banskota AH, Tran QK, Saiki I, Kadota S (2001) New Spirostanol Steroids and Steroidal Saponins from Roots and Rhizomes of *Dracaena angustifolia* and Their Antiproliferative Activity. J Nat Prod **64**: 1127
197. Rubinstein LV, Shoemaker RH, Paull KD, Simon RM, Tosini S, Skehan P, Scudiero DA, Monks A, Boyd MR (1990) Comparison of *in vitro* Anticancer-Drug-Screening Data Generated with a Tetrazolium Assay *versus* a Protein Assay Against a Diverse Panel of Human Tumor Cell Lines. J Nat Cancer Inst **82**: 1113
198. Dimoglo AS, Choban IN, Bersuker IB, Kintya PK, Balashova NN (1985) Structure-Activity Correlations for the Antioxidant and Antifungal Properties of Steroid Glycosides. Bioorg Khim **11**: 408

199. Imai S, Fujioka S, Murata E, Goto M, Kawasaki T, Yamauchi T (1967) Bioassay of Crude Drugs and Oriental Crude Drug Preparations. XXII. Search for Biologically Active Plant Ingredients by Means of Antimicrobial Tests. 4. Antifungal Activity of Dioscin and Related Compounds. Takeda Kenkyusho Nenpo **26**: 76
200. Wolters B (1965) The Share of the Steroid Saponins in the Antibiotic Action of *Solanum dulcumara*. Planta Med **13**: 189
201. Wolters B (1966) Antimicrobial Activity of Plant Steroids and Triterpenes. Planta Med **14**: 392
202. Chen H, Xu Y, Jiang Y, Wen H, Cao Y, Liu W, Zhang J (2003) Application of *Tribulus terrestris* Spirosteroidal Saponin to Prepare the Antifungal Medical Preparations. Faming Zhuanli Shenqing Gongkai Shuomingshu. China Patent 1428349
203. De Lucca AJ, Bland JM, Vigo CB, Selitrennikoff MCP (2001) Fungicidal Saponin, CAY-1, and Isolation Thereof from *Capsicum* Species Fruit. US Patent 6,310,091
204. Magota H, Okubo K, Shimoyamada M, Suzuki M, Maruyama M (1991) Isolation of Steroidal Saponin as Antifungal Agent. Japan Patent 03048694
205. Sashida Y, Mitsumaki Y, Kuroda A, Takashi T, Sudo K (2001) Antifungal Steroid Saponin. Japan Patent 2001181296
206. Yang C-R, Zhang Y, Jacob MR, Khan SI, Zhang Y-J, Li Z-C (2006) Antifungal Activity of C-27 Steroidal Saponins. Antimicrob Agents Chemother **50**: 1710
207. Zhang Y, Li H-Z, Zhang Y-J, Jacob MR, Khan SI, Li X-C, Yang C-R (2006) Atropurosides A–G, New Steroidal Saponins from *Smilacina atropurpurea*. Steroids **71**: 712
208. NCCLS (2002) Reference Method for Broth Dilution Antifungal Susceptibility Testing of Yeasts: Approved Standard M-27-A2, 22 (15). National Committee on Clinical Laboratory Standards, Wayne, PA
209. NCCLS (2002) Reference Method for Broth Dilution Antifungal Susceptibility Testing of Yeasts: Approved Standard M-38-A, 22 (16). National Committee on Clinical Laboratory Standards, Wayne, PA
210. Favel LA, Kemertelidze E, Benidze M, Fallague K, Regli P (2005) Antifungal Activity of Steroidal Glycosides from *Yucca gloriosa*. Phytother Res **19**: 158
211. Miyakoshi M, Tamura Y, Masuda H, Mizutani K, Tanaka O, Ikeda T, Ohtani K, Kasai R, Yamasaki K (2000) Antiyeast Steroidal Saponins from *Yucca schidigera* (Mohave Yucca), a New Anti-Food-Deteriorating Agent. J Nat Prod **63**: 332
212. Rahalison L, Hamburger M, Monod M, Hostettmann K (1994) Antifungal Tests in Phytochemical Investigations: Comparison of Bioautographic Methods Using Phytopathogenic and Human Pathogenic Fungi. Planta Med **60**: 41
213. González M, Zamilpa A, Marquina S, Navarro V, Alvarez L (2004) Antimycotic Spirostanol Saponins from *Solanum hispidum* Leaves and Their Structure-Activity Relationships. J Nat Prod **67**: 938
214. Zamilpa A, Tortoriello J, Navarro V, Delgado G, Alvarez L (2002) Five New Steroidal Saponins from *Solanum chrysotrichum* Leaves and Their Antimycotic Activity. J Nat Prod **65**: 1815
215. Sautour M, Mitaine-Offer A-C, Miyamoto T, Dongmo A, Lacaille-Dubois MA (2004) Antifungal Steroid Saponins from *Dioscorea cayenensis*. Planta Med **70**: 90
216. Sautour M, Mitaine-Offer A-C, Miyamoto T, Dongmo A, Lacaille-Dubois MA (2004) A New Steroidal Saponin from *Dioscorea cayenensis*. Chem Pharm Bull **52**: 1353
217. Takechi M, Tanaka Y (1991) Structure-Activity Relationships of Synthetic Diosgenyl Monoglycosides. Phytochemistry **30**: 2557

218. Lacaille-Dubois MA, Wagner H (1996) A Review of the Biological and Pharmacological Activities of Saponins. Phytomedicine **2**: 363
219. Santos WR, Bernardo RR, Pecanha LMT, Palatnik M, Parente JP, De Sousa CBP (1997) Haemolytic Activities of Plant Saponins and Adjuvants. Effect of *Periandra mediterranea* Saponin on the Humoral Response to the FML Antigen of *Leishmania donovani*. Vaccine **15**: 1024
220. Mendes TP, Silva GDM, Silva BPD, Parente JE (2004) A New Steroidal Saponin from *Agave attenuata*. Nat Prod Res **18**: 183
221. Harris KF (1977) An Ingestion-Egestion Hypothesis of Noncirculative Virus Transmission. In: Harris KF, Maramorosch K (eds.) Aphids as Virus Vectors, p. 165. Academic Press, New York
222. Raman KV, Radcliffe EB (1992) Pest Aspects of Potato Productions. In: Harris P (ed.) The Potato Crop, 2nd Edn., p. 477. Chapman & Hall, London
223. Soulé S, Güntner C, Vázquez A, Argandona V, Moyna P, Ferreira F (2000) An Aphid Repellent Glycoside from *Solanum laxum*. Phytochemistry **55**: 217
224. Harwood HJ, Chandler CE, Pellatin LD, Bangerter FW, Wilkins RW, Long CA, Cosgrove PG, Malinow MR, Marzetta CA, Pettini JL, Savoy YE, Mayne JT (1993) Pharmacologic Consequences of Cholesterol Absorption Inhibition: Alteration in Cholesterol Metabolism and Reduction in Plasma Cholesterol Concentration Induced by the Synthetic Saponin β-Tigogenin Cellobioside (CP-88818; Tiqueside). J Lipid Res **34**: 377
225. Koch HP (1993) Saponine in Knoblauch und Küchenzwiebel. Dtsch Apoth Ztg **133**: 3733
226. Matsuura H (2001) Saponins in Garlic as Modifiers of the Risk of Cardiovascular Disease. J Nutr **131**: 1000S
227. Dutta A, Mandal D, Mondal NB, Banerjee S, Sahu NP, Mandal C (unpublished results)
228. Wink M (1999) Functions of Plant Secondary Metabolites and Their Exploitation in Biotechnology. Sheffield Academic Press, Sheffield
229. Tschesche R (1971) Advances in the Chemistry of Antibiotic Substances from Higher Plants. In: Wagner H, Hörhammer I (eds.) Pharmacognosy and Phytochemistry, p. 274. Springer, Berlin Heidelberg New York
230. Schönbeck F, Schlösser E (1976) Preformed Substances as Potential Phytoprotectants. In: Heitefuss R, Williams PH (eds.) Physiological Plant Pathology, p. 653. Springer, Berlin Heidelberg New York
231. Osbourn AE (1996) Pre-Formed Antimicrobial Compounds and Plant Defence Against Fungal Attack. Plant Cell **8**: 1821
232. Tschesche R (1972) Biosynthesis of Cardinolides, Bufadienolides and Steroid Sapogenins. Proc Royal Soc (B) **180**: 187
233. Heftmann E (1983) Biogenesis of Steroids in Solanaceae. Phytochemistry **22**: 1843
234. Tal B, Tamir J, Rokem JS, Goldberg I (1984) Isolation and Characterization of an Intermediate Steroid Metabolite in Diosgenin Biosynthesis in Suspension Cultures of *Dioscorea deltoidea* Cells. Biochem J **219**: 619
235. Gurielidze KG, Pasehnichenko VA, Vasil'eva IS (1987) Glucohydrolase from the Leaves and Roots of *Dioscorea deltoidea* Wall. Biokhimia **52**: 362
236. Kalinowska M, Wojciechowski ZA (1986) Enzymatic Synthesis of Nuatigenin 3β-D-Glucoside in Oat (*Avena sativa* L.) Leaves. Phytochemistry **25**: 2525
237. Kalinowska M, Wojciechowski ZA (1987) Subcellular Localization of UDPG: Nuatigenin Glucosyltransferase in Oat Leaves. Phytochemistry **26**: 353

238. Paczkowski C, Zimowski J, Krawczyk D, Wojciechowski ZA (1990) Steroid-Specific Glucosyltransferases in *Asparagus plumosus* Shoots. Phytochemistry **29**: 63
239. Indrayanto G, Zumaroh S, Syahrani A, Wilkins AL (2001) C-27 and C-3 Glucosylation of Diosgenin by Cell Suspension Cultures of *Costus speciosus*. J Asian Nat Prod Res **3**: 161
240. Kalinowska M, Zimowski J, Paczkowski C, Wojciechowski ZA (2005) The Formation of Sugar Chains in Triterpenoid Saponins and Glycoalkaloids. Phytochem Rev **4**: 237
241. Jin J-M, Zhang Y-J, Yang C-R (2004) Four New Steroid Constituents from the Waste Residue of Fibre Separation from *Agave americana* Leaves. Chem Pharm Bull **52**: 654
242. Da Silva BP, De Sousa AC, Silva GM, Mendes TP, Parente JP (2002) A New Bioactive Steroidal Saponin from *Agave attenuata*. Z Naturforsch **57c**: 423
243. Silva GM, De Souza AM, Lara LS, Mendes TP, Da Silva BP, Lopes AG, Caruso-Neves C, Parente JP (2005) A New Steroidal Saponin from *Agave brittoniana* and Its Biphasic Effect on the Na^+-ATPase Activity. Z Naturforsch **60c**: 121
244. Abdel-Gawad MM, El-Sayed MM, Abdel-Hameed ES (1999) Molluscicidal Steroidal Saponins and Lipid Content of *Agave decipiens*. Fitoterapia **70**: 371
245. Da Silva BP, Parente JP (2005) A New Bioactive Steroidal Saponin from *Agave shrevei*. Z Naturforsch **60c**: 57
246. Barile E, Zolfaghari B, Sajjadi SE, Lanzotti V (2004) Saponins of *Allium elburzense*. J Nat Prod **67**: 2037
247. Akhov LS, Musienko MM, Piacente S, Pizza C, Oleszek W (1999) Structure of Steroidal Saponins from Underground Parts of *Allium nutans* L. J Agric Food Chem **47**: 3193
248. Carotenuto A, Fattorusso E, Lanzotti V, Magno S (1999) Spirostanol Saponins from *Allium porrum* L. Phytochemistry **51**: 1077
249. Zou Z-M, Yu D-Q, Cong P-Z (2001) A Steroidal Saponin from the Seeds of *Allium tuberosum*. Phytochemistry **57**: 1219
250. Ikeda T, Tsumagari H, Okawa M, Nohara T (2004) Pregnane- and Furostane-Type Oligoglycosides from the Seeds of *Allium tuberosum*. Chem Pharm Bull **52**: 142
251. Zhang H-J, Sydara K, Tan GT, Ma C, Southavong B, Soejarto DD, Pezzuto JM, Fong HHS (2004) Bioactive Constituents from *Asparagus cochinchinensis*. J Nat Prod **67**: 194
252. Li Y-F, Hu L-H, Lou F-C, Hong J-R, Li J, Shen Q (2005) Furostanoside from *Asparagus filicinus*. J Asian Nat Prod Res **7**: 43
253. Huang X, Kong L (2006) Steroidal Saponins from Roots of *Asparagus officinalis*. Steroids **71**: 171
254. Farid H, Haslinger E, Kunert O, Wegner C, Hamburger M (2002) Steroidal Glycosides from *Balanites aegyptiaca*. Helv Chim Acta **85**: 1019
255. Kuroda M, Mimaki Y, Hasegawa F, Yokosuka A, Sashida Y, Sakagami H (1999) Steroidal Glycosides from the Bulbs of *Camassia leichtlinii* and Their Cytotoxic Activities. Chem Pharm Bull **47**: 738
256. Haraguchi M, Motidome M, Morita H, Takeya K, Itokawa H, Mimaki Y, Sashida Y (1999) New Polyhydroxylated Steroidal Sapogenin and Saponin from the Leaves of *Cestrum sendtenerianum*. Chem Pharm Bull **47**: 582
257. Mimaki Y, Kuroda M, Takaashi Y, Sashida Y (1998) Steroidal Saponins from the Leaves of *Cordyline stricta*. Phytochemistry **47**: 79
258. Da Silva BP, Bernardo RR, Parente JP (1999) A Furostanol Glycoside from Rhizomes of *Costus spicatus*. Phytochemistry **51**: 931

259. Dong M, Feng X-Z, Wang B-X, Wu L-J, Ikejima T (2001) Two Novel Furostanol Saponins from the Rhizomes of *Dioscorea panthaica* Prain et Burkill and Their Cytotoxic Activity. Tetrahedron **57**: 501
260. Dong M, Feng X-Z, Wu L-J, Wang B-X, Ikejima T (2001) Two New Steroidal Saponins from the Rhizomes of *Dioscorea panthaica* and Their Cytotoxic Activity. Planta Med **67**: 853
261. Osorio JN, Martinez OMM, Navarro YMC, Watanabe K, Sakagami H, Mimaki Y (2005) Polyhydroxylated Spirostanol Saponins from the Tubers of *Dioscorea polygonoides*. J Nat Prod **68**: 1116
262. Yang D-J, Lu T-J, Hwang LS (2003) Isolation and Identification of Steroidal Saponins in Taiwanese Yam Cultivar (*Dioscorea pseudojaponica* Yamamoto). J Agric Food Chem **51**: 6438
263. Yang Q-X, Xu M, Zhang Y-J, Li H-Z, Yang C-R (2004) Steroidal Saponins from *Disporopsis pernyi*. Helv Chim Acta **87**: 1248
264. Zheng Q-A, Zhang Y-J, Li H-Z, Yang C-R (2004) Steroidal Saponins from Fresh Stem of *Dracaena cochinchinensis*. Steroids **69**: 111
265. Mimaki Y, Kuroda M, Takaashi Y, Sashida Y (1998) Steroidal Saponins from the Stems of *Dracaena concinna*. Phytochemistry **47**: 1351
266. Gonzalez AG, Hernandez JC, Leon F, Padron JI, Estevez F, Quintana J, Bermejo J (2003) Steroidal Saponins from the Bark of *Dracaena draco* and Their Cytotoxic Activities. J Nat Prod **66**: 793
267. Yokosuka A, Mimaki Y, Sashida Y (2000) Steroidal Saponins from *Dracaena surculosa*. J Nat Prod **63**: 1239
268. Joanne L, Boyce S, Tinto WF, McLean S, Reynolds WF (2004) Saponins from *Furcraea selloa* var. *marginata*. Fitoterapia **75**: 634
269. He X, Qiu F, Shoyama Y, Tanaka H, Yao X (2002) Two New Steroidal Saponins from "Gualou-xiebai-baijiu-tang" Consisting of *Fructus Trichosanthis* and *Bulbus Allii Macrostemi*. Chem Pharm Bull **50**: 653
270. Watanabe K, Mimaki Y, Sakagami H, Sashida Y (2003) Bufadienolide and Spirostanol Glycosides from the Rhizomes of *Helleborus orientalis*. J Nat Prod **66**: 236
271. Braca A, Prieto JM, De Tommasi N, Tomè F, Morelli I (2004) Furostanol Saponins and Quercetin Glycosides from the Leaves of *Helleborus viridis* L. Phytochemistry **65**: 2921
272. Konishi T, Fujiwara Y, Konoshima T, Kiyosawa S, Nishi M, Miyahara K (2001) Steroidal Saponins from *Hemerocallis fulva* var. *kwanso*. Chem Pharm Bull **49**: 318
273. Mimaki Y, Satou T, Kuroda M, Sashida Y, Hatakeyama Y (1999) Steroidal Saponins from the Bulbs of *Lilium candidum*. Phytochemistry **51**: 567
274. Mimaki Y, Satou T, Kuroda M, Sashida Y, Hatakeyama Y (1998) New Steroidal Constituents from the Bulbs of *Lilium candidum*. Chem Pharm Bull **46**: 1829
275. Dai H-F, Deng S-M, Tan N-H, Zhou J (2005) A New Steroidal Glycoside from *Ophiopogon japonicus* (Thunb.) Ker-Gawl. J Integrative Plant Biol **47**: 1148
276. Kuroda M, Mimaki Y, Ori K, Sakagami H, Sashida Y (2004) Steroidal Glycosides from the Bulbs of *Ornithogalum thyrsoides*. J Nat Prod **67**: 1690
277. Kuroda M, Ori K, Mimaki Y (2006) Ornithosaponins A–D, Four New Polyoxygenated Steroidal Glycosides from the Bulbs of *Ornithogalum thyrsoides*. Steroids **71**: 199
278. Jin J-M, Zhang Y-J, Li H-Z, Yang C-R (2004) Cytotoxic Steroidal Saponins from *Polygonatum zanlanscianense*. J Nat Prod **67**: 1992

279. Mimaki Y, Kuroda M, Yokosuka A, Sashida Y (1998) Two New Bisdesmosidic Steroidal Saponins from the Underground Parts of *Ruscus aculeatus*. Chem Pharm Bull **46**: 879
280. Mimaki Y, Kuroda M, Yokosuka A, Sashida Y (1999) A Spirostanol Saponin from the Underground Parts of *Ruscus aculeatus*. Phytochemistry **51**: 689
281. Honbu T, Ikeda T, Zhu X-H, Yoshihara O, Okawa M, Nafady AM, Nohara T (2002) New Steroidal Glycosides from the Fruits of *Solanum anguivi*. J Nat Prod **65**: 1918
282. Zhu X-H, Ikeda T, Nohara T (2000) Studies on Constituents of Solanaceous Plants. (46). Steroidal Glycosides from the Fruits of *Solanum anguivi*. Chem Pharm Bull **48**: 568
283. Putalun W, Xuan L-J, Tanaka H, Shoyama Y (1999) Solakhasoside, A Novel Steroidal Saponin from *Solanum khasianum*. J Nat Prod **62**: 181
284. Ferro EA, Alvarenga NL, Ibarrola DA, Hellion-Ibarrola MC, Ravelo AG (2005) A New Steroidal Saponin from *Solanum sisymbriifolium* Roots. Fitoterapia **76**: 577
285. Ono M, Nishimura K, Suzuki K, Fukushima T, Igoshi K, Yoshimitsu H, Ikeda T, Nohara T (2006) Steroidal Glycosides from the Underground Parts of *Solanum sodomaeum*. Chem Pharm Bull **54**: 230
286. Iida Y, Yanai Y, Ono M, Ikeda T, Nohara T (1998) Three Unusual 22-β-O-23-Hydroxy-(5α)-Spirostanol Glycosides from the Fruits of *Solanum torvum*. Chem Pharm Bull **53**: 1112
287. Temraz A, El Gindi OD, Kadry HA, De Tommasi N, Braca A (2006) Steroidal Saponins from the Aerial Parts of *Tribulus alatus* Del. Phytochemistry **67**: 1011
288. Perrone A, Plaza A, Bloise E, Nigro P, Hamed AI, Belisario MA, Pizza C, Piacente S (2005) Cytotoxic Furostanol Saponins and a Megastigmane Glucoside from *Tribulus parvispinus*. J Nat Prod **68**: 1549
289. Combarieu ED, Fuzzati N, Lovati M, Mercalli E (2003) Furostanol Saponins from *Tribulus terrestris*. Fitoterapia **74**: 583
290. Xu Y-X, Chen H-S, Liang H-Q, Gu Z-B, Liu W-Y, Leung W-N, Li T-J (2000) Three New Saponins from *Tribulus terrestris*. Planta Med **66**: 545
291. Kostova I, Dinchev D, Rentsch GH, Dimitrov V, Ivanova A (2002) Two New Sulfated Furostanol Saponins from *Tribulus terrestris*. Z Naturforsch **57c**: 33
292. Huang J-W, Tan C-H, Jiang S-H, Zhu D-Y (2003) Terrestrinins A and B, Two New Steroid Saponins from *Tribulus terrestris*. J Asian Nat Prod Res **5**: 285
293. Saxena VK, Shalem A (2004) Yamogenin-3-O-β-D-glucopyranosyl-(1 $\rightarrow$ 4)-O-α-D-xylopyranoside from the Seeds of *Trigonella foenum-graecum*. J Chem Sci **116**: 79
294. Murakami T, Kishi A, Matsuda H, Yoshikawa M (2000) Medicinal Foodstuffs. XVII. Fenugreek Seed. (3): Structures of New Furostanol Type Steroid Saponins, Trigoneosides Xa, Xb, XIb, XIIa, XIIb and XIIIa, from the Seeds of Egyptian *Trigonella foenum-graecum* L. Chem Pharm Bull **48**: 994
295. Shen P, Wang S-L, Liu X-K, Yang C-R, Cai B, Yao X-S (2003) Steroidal Saponins from Rhizomes of *Tupistra wattii* Hook. f. Chem Pharm Bull **51**: 305
296. Yang Q-X, Zhang Y-J, Li H-Z, Yang C-R (2005) Polyhydroxylated Steroidal Constituents from the Fresh Rhizomes of *Tupistra yunnanensis*. Steroids **70**: 732
297. Ozipek M, Saracoglu I, Ogihara Y, Calii S (2002) Nuatigenin-Type Steroidal Saponins from *Veronica fuhsii* and *V. multifida*. Z Naturforsch **57c**: 603
298. Oleszek W, Sitek M, Stochmal A, Piacente S, Pizza C, Cheeke P (2001) Steroidal Saponins of *Yucca schidigera* Roezl. J Agric Food Chem **49**: 4392

Author Index

Page numbers printed in *italics* refer to References

Abdel-Gawad, M.M. *139*
Abdel-Hameed, E.S. *139*
Abdullaev, N.D. *130*
Abebe, D. *131*
Abliz, Z. *133*
Aboutabl, E.A. *129*
Abubakirov, N.K. *130*
Achari, B. *129, 134*
Affar, E.B. *136*
Agrawal, P.K. 58, *128, 133*
Ahn, K.-J. *136*
Akhov, L.S. *139*
Akiyama, T. *130*
Aksyonov, S.A. *132*
Alvarenga, N.L. *141*
Alvarez, L. *137*
Ames, B.N. 43
Amoros, M. *129*
Amrein, W. 38
Anderl, M. *41*
Anders, I. 40, *41*
Anderson, N.L. *135*
Ando, Y. *135*
Aquino, R. *131*
Argandona, V. *138*
Arita, H. *131*
Armstead, I. 38
Arthan, D. *130*
Asakawa, J. *131*
Aubry, S. 38, 40, *41*
Auclair, K. 38
Ausubel, F.M. *41*
Authi, K.S. *134*

Bachmann, A. 39
Bai, Y. *132*
Baissac, Y. *127*
Bakinovskii, L.B. *130*
Balashova, N.N. *136*
Banerjee, S. *128, 129, 138*
Bang, J.-K. *136*

Bangerter, F.W. *138*
Banskota, A.H. *136*
Barile, E. *139*
Barr, I.G. *135*
Bates, S.E. *135*
Bax, A. *133, 134*
Beale, S.I. 38, 42
Becker, K. *127, 135*
Becker, R.C. *129*
Bedir, E. *133*
Belisario, M.A. *141*
Bendall, M.R. *133*
Benecke, I. *131*
Benidze, M. *137*
Benie, T. *134*
Berghold, J. 40, 42
Bermejo, J. *136, 140*
Bernardo, R.R. *138, 139*
Bersuker, I.B. *136*
Beyer, G. *130*
Bianchi, E. *129*
Bisset, G.M.F. 38
Bland, J.M. *137*
Bloise, E. *141*
Bock, K. *134*
Bodenhausen, G. *134*
Boelens, R. *134*
Bojesen, G. *131*
Bollivar, D.W. 38
Bortlik, K. 38, 40
Bothner-By, A.A. *134*
Bouwkamp, J.C. 39
Bovet, L. *41*
Boyce, S. *140*
Boyd, M.R. *135, 136*
Braca, A. *140, 141*
Bradley, G. *131*
Branaman, J.R. *129*
Brattain, D.E. *136*
Brattain, M.G. *136*
Braunschweiler, L. *133*

Breit, S. *132*
Brereton, R.G. *42*
Bretting, H. *131*
Breuker, K. *40*
Brobera, S. *133*
Bross-Walch, N. 60, *134*
Brown, S.B. *37, 38*
Brunner, H. *130*
Bunow, B. *135*
Bunsawansong, P. *133*
Buolamwini, J.K. *135*

Cai, B. *141*
Calii, S. *141*
Callot, H.J. *43*
Campbell, H. *135*
Cao, Y. *137*
Carlson, R.M.K. 60, *134*
Carmichael, J. *136*
Carotenuto, A. *139*
Caruso-Neves, C. *139*
Castillo, A.R. *41*
Chai, W. *132*
Chakravarti, R.N. *128*
Chakravarty, A.K. *128*
Chan, Z. *133*
Chandler, C.E. *138*
Chang, M. *131*
Chapuis, J.-C. *128*
Cheeke, P. *141*
Chen, H. *137*
Chen, H.-S. *141*
Cheong, J.H. *136*
Chin, Y.-W. *136*
Choban, I.N. *136*
Claridge, T.D.W. *134*
Cole, J.R. *129*
Combarieu, E.D. *141*
Cong, P.-Z. *139*
Cornejo, J. *42*
Cosgrove, P.G. *138*
Cox, J.C. *135*
Crawford, N. *134*
Croasmun, W.R. 60, *134*
Cronise, P. *135*
Cui, M. *132*
Curty, C. *40, 41, 43*

Dai, H.-F. *140*
D'Amours, D. *136*

Dangl, J.L. *37*
Da Silva, B.P. *139*
D'Auria, M.V. *134*
David, Z.E. *136*
Davis, D.G. *133, 134*
Debella, A. *131*
Deditus, C. *134*
Degraff, W.G. *136*
Delgado, G. *137*
De Lucca, A.J. *137*
Deng, S.-M. *140*
De Simone, F. *131*
Desnoyers, S. *136*
De Sousa, A.C. *139*
De Sousa, C.B.P. *138*
De Souza, A.M. *139*
De Tommasi, N. *140, 141*
DeWarrd, P. *134*
Dietrich, R.A. *37*
Dimitrov, V. *141*
Dimoglo, A.S. *136*
Dinchev, D. *141*
Ding, L. *133*
Dini, A. *131*
Dixit, V.M. *136*
Djerassi, C. *127, 131*
Do, J.C. *130*
Doddwell, D.M. *133*
Doi, M. *39*
Dong, F. *128*
Dong, M. *140*
Dongmo, A. *137*
Donnison, I. *38, 41*
Doubek, D.L. *128*
Düggelin, T. *39, 40*
Dunlap, J.C. *43*
Dutta, A. *138*
Duval, J. *134*

Ebizuka, Y. *130*
Eggert, H. *131*
Eichhorn, S.E. *43*
Eichmüller, C. *40*
El Gindi, O.D. *141*
El Izzi, A. *134*
Elks, J. 46, *127*
El-Sayed, M.M. *139*
Engel, N. *39–43*
Ernst, R.R. *133, 134*
Eschenmoser, A. *42*

Esen, A. *130*
Estévez, F. *136, 140*
Evenden, B.J. *134*
Evert, R.F. *43*
Eyam, Y. *43*

Falk, H. *42*
Fallague, K. *137*
Fang, S.P. *132*
Farid, H. *139*
Fattorusso, E. *139*
Favel, L.A. *137*
Feng, X.-Z. *140*
Fenselau, C. *131*
Fernandez-Lopez, J. *39*
Ferreira, F. *138*
Ferro, E.A. *141*
Fieser, L. *128*
Fieser, M. *128*
Fincher, G.B. *130*
Fine, W.D. *136*
Fohlman, J. *132*
Fojo, T. *135*
Folly, P. *39*
Fong, H.H.S. *139*
Fornace, A.J. Jr. *135*
Fotsis, T. *132*
Francis, G. *127, 135*
Frankenberg, N. *42*
Fridriksson, E.K. *132*
Friend, S.H. *135*
Frydman, B. *39*
Fujioka, S. *137*
Fujiwara, Y. *140*
Fukasawa, T. *127, 130, 135*
Fukuda, N. *129*
Fukushima, T. *141*
Furuya, S. *135*
Fusetani, N. *136*
Fuzzati, N. *141*

Ganguly, A.N. *127, 129, 131*
Gao, H. *136*
Gazdar, A.F. *136*
Gehrke, M. *133*
George, A.J. *134*
Gerlach, B. *41*
Germain, M. *136*
Giannini, C. *134*

Ginsburg, S. *38–40, 42*
Girre, R.L. *129*
Glasenapp-Breiling, M. *38*
Glaser, J. *133*
Glazer, A.N. *43*
Goda, Y. *128, 133*
Goldberg, I. *138*
Goldschmidt, E.E. *43*
Goldstein, I.J. *130*
Gomita, Y. *129*
Gonzalez, A.G. *140*
González, M. *137*
Gorovits, M.B. *130*
Gossauer, A. 18, 30, *40, 41, 43*
Goto, M. *137*
Gray-Goodrich, M. *135*
Greenberg, J.T. *41*
Griesinger, C. *133*
Grimm, B. *38*
Gründemann, E. *130*
Gu, Z.-B. *141*
Guillaume, D. *131*
Güntner, C. *138*
Guo, A.L. *41*
Guo, M.Q. *132*
Gurielidze, K.G. *138*
Gus-Mayer, S. *130*
Guy, R.R. *136*

Hakansson, P. *132*
Hamada, T. *129*
Hamburger, M. *137, 139*
Hamed, A.I. *141*
Hao, C.Y. *132*
Haraguchi, M. *133, 139*
Harborne, J.B. *42*
Hardman, R. *127*
Harper, J. *38*
Harris, K.F. *138*
Harvey, A.J. *130*
Harwood, H.J. *138*
Hase, E. *42*
Hasegawa, F. *139*
Haslinger, E. *131, 139*
Hastings, J.W. *38, 43*
Hatakeyama, Y. *140*
Hauck, P.R. *127*
Hay, G.W. *130*
Hayashi, K. *130*
Hayata, Y. *136*

He, X. *136, 140*
Hedin, A. *132*
Heftmann, E. *138*
Hellion-Ibarrola, M.C. *141*
Hendry, G.A.F. *37*
Hernández, J.C. *66, 136, 140*
Hillenkamp, F *132*
Hiller, K. *127, 130*
Hinder, B. *42*
Ho, C.-T. *133*
Hodes, L. *135*
Hoffmann, H. *128*
Hoj, P.B. *130*
Holland, D. *43*
Honbu, T. *136, 141*
Hong, J. *127*
Hong, J.-R. *139*
Hoogstraten, R. *41*
Horn, D.M. *132*
Hörtensteiner, S. *37–43*
Hose, C. *135*
Hostettmann, K. *127, 131, 137*
Houghton, J.D. *37*
Houthaeve, T. *132*
Hrmova, M. *130*
Hu, K. *135*
Hu, L.-H. *139*
Huang, J.-W. *141*
Huang, X. *139*
Huang, Y. *134*
Hughes, M.A. *129*
Hui, Y.Z. *128*
Hunziker, P.E. *41*
Hwang, L.S. *140*
Hynninen, P.H. *39*

Ibarrola, D.A. *141*
Ide, A. *135*
Igoshi, K. *141*
Iida, I. *130*
Iida, Y. *141*
Ikeda, T. *65, 127, 129, 136, 137, 139, 141*
Ikejima, T. *140*
Imai, S. *137*
Imamura, N. *129*
Inage, T. *39*
Indrayanto, G. *139*
Inoue, K. *130, 131*
Ishibashi, M. *128, 133*
Ishihara, A. *129*

Ito, H. *39*
Ito, Y. *129*
Itokawa, H. *133, 139*
Iturraspe, J. *39, 41, 43*
Ivanova, A. *141*
Iwamatsu, A. *43*
Iwamura, H. *129*

Jacob, M.R. *137*
Jacob-Wilk, D. *43*
Jain, D.C. *128*
James, C. *38*
Jaun, B. *38*
Jeanloz, R.W. *134*
Jenny, T.A. *40*
Jia, Z. *127, 133, 134*
Jiang, S.-H. *141*
Jiang, Y. *137*
Jin, J.-M. *133, 139, 140*
Joanne, L. *140*
Johnson, G.S. *135*
Jones, G.P. *130*

Kadota, S. *136*
Kadry, H.A. *141*
Kaku, H. *130*
Kalinkevich, A.N. *132*
Kalinowska, M. *138, 139*
Kamensky, M. *132*
Kamerling, J.P. *132*
Kameyama, A. *129, 135*
Kaneko, K. *130*
Kaneko, Y. *130*
Kang, L. *128*
Kanmoto, T. *129*
Karas, M. *132*
Kasai, R. *127, 130, 131, 137*
Kasai, Y.R. *131*
Kashibuchi, N. *127*
Kawahara, N. *128, 133*
Kawasaki, T. *128, 129, 131, 132, 135, 137*
Keller, F. *38*
Kemertelidze, E. *127, 137*
Kenne, L. *133*
Kensil, C.R. *135*
Kerem, Z. *127, 135*
Kessler, H. *133*
Khaled, F.M. *136*
Khan, I.A. *133*

Author Index

Khan, S.I. *137*
Khon, K.W. *135*
Kim, C.Y. *136*
Kim, G.-S. *136*
Kim, H.-T. *136*
Kim, J. *136*
King, I. *38*
Kinjo, M. *136*
Kintia, P.K. *128*, *132*
Kintya, P.K. *129*, *130*, *136*
Kishi, A. *141*
Kishi, Y. 5, 32, 33, *38*, *41*, *43*
Kiss, L. *129*
Kitada, Y. *127*
Kittakoop, P. *130*
Kiyosawa, S. *140*
Klessig, D.F. *41*
Klyne, W. 52, *131*
Knight, J.C. *128*
Koch, H.P. 68, *138*
Kohchi, T. *42*
Kohga, S. *136*
Koike, K. *129*, *133*, *134*
Komori, T. *128*, *129*, *132*
Kong, L. *139*
König, W.A. *131*
Konishi, T. *140*
Konoshima, T. *140*
Kostova, I. *141*
Kowithayakorn, T. *128*, *133*
Koyano, T. *128*, *133*
Krause, E. *130*
Kräutler, B. *37–42*
Krawczyk, D. *139*
Křen, V. *127*, *135*
Krider, M.M. *129*
Krokhmalyuk, V.V. *129*, *130*
Kudou, S. *135*
Kühn, T. *134*
Kunert, O. *131*, *139*
Kuroda, A. *137*
Kuroda, M. *127*, *129*, *130*, *135*, *139–141*

Lacaille-Dubois, M.A. *128*, *137*, *138*
Lagarias, J.C. *42*
Langley, J. *135*
Langmeier, M. *39*
Lanzotti, V. *139*
Lao, A. *133*
Lara, L.S. *139*

Lattimer, R.P. *132*
Lavaud, C. *131*
Lawson, A.M. *132*
Leconte, O. *127*
Lee, C.-O. *136*
Lee, J. *134*
Leeflang, B.R. *132*
Le Men-Olivier, L. *131*, *134*
León, F. *136*, *140*
Leopold, A.C. *43*
Leung, W.-N. *141*
Lewis, B.A. *130*
Lewis, M.A. *132*
Li, H.-Z. *137*, *140*, *141*
Li, J. *139*
Li, L.-J. *133*
Li, R. 55, *133*
Li, T.-J. *141*
Li, X.-C. *128*, *130*, *137*
Li, Y.-F. *139*
Li, Z.-C. *137*
Liang, F. 56, *133*
Liang, H.-Q. *141*
Liljegren, S.J. *40*
Lindberg, M. *132*
Liu, S.Y. *132*
Liu, W. *137*
Liu, W.-Y. *141*
Liu, X.-K. *141*
Liu, Z.Q. *132*
Llewellyn, C.A. *42*
Long, C.A. *138*
Lopes, A.G. *139*
Lopez, J. 49, *128*
Losey, F.G. *42*
Lou, F.-C. *139*
Lovati, M. *141*
Lu, T.-J. *140*

Ma, B. *128*
Ma, C. *139*
Mach, J.M. *41*
Mackie, A.M. *130*
Macura, S. *134*
Magno, S. *139*
Magota, H. *137*
Mahato, S.B. *127–129*, *131*
Makkar, H.P.S. *127*, *135*
Malinow, M.R. *138*
Mandal, C. *138*

Author Index

Mandal, D. *128, 138*
Mani, J. *38*
Mann, M. *132*
Mantoura, R.F.C. *42*
Mao, S. *133*
Marker, R.E. 49, *128*
Marquina, S. *137*
Marston, A. *127*
Martinez, O.M.M. *140*
Martinková, L. *127, 135*
Martinoia, E. *42*
Maruyama, M. *137*
Marx, R.S. *133*
Marzetta, C.A. *138*
Marzilli, L.G. *134*
Massiot, G. *131*
Masuda, H. *127, 137*
Masuda, T. *43*
Matile, P. 5, *37–43*
Matsubara, K. *127*
Matsuda, H. *141*
Matsunaga, S. *136*
Matsuura, H. *130, 138*
Mayne, J.T. *138*
Mayo, J. *135*
McDonagh, A.F. *43*
McLafferty, F.W. *132*
McLean, S. *140*
Mendel, G. 6, 36, *38*
Mendes, T.P. *138, 139*
Mercalli, E. *141*
Meselhy, M.R. *129*
Meyers, H.V. *131*
Michi, C. *131*
Mikaki, Y. *131*
Mimaki, Y. 51, 63, 64, *127–130, 133, 135, 136, 139–141*
Minale, L. *134*
Minna, J.D. *136*
Mirkin, G. *127*
Mitaine-Offer, A.-C. *137*
Mitchell, J.B. *136*
Mitsuhashi, H. *130*
Mitsumaki, Y. *137*
Miyahara, K. *128, 131, 135, 140*
Miyakoshi, M. 67, *127, 137*
Miyamoto, T. *128, 137*
Mizutani, K. *127, 131, 137*
Moffet, M. *38*
Mondal, N.B. *128, 138*

Monks, A. *135, 136*
Monod, M. *137*
Montforts, F.P. *38*
Mooser, V. *40*
Morel, A. *43*
Morelli, I. *140*
Morita, H. *133, 139*
Moriyama, M. *129*
Morris, G.A. *133*
Morse, D. 38, *41*
Moser, S. 39, *40*
Moskau, D. *134*
Mosmann, T. *136*
Motidome, M. *133, 139*
Moyano, N. *39*
Moyna, P. *138*
Mukougawa, K. *42*
Mühlecker, W. *39–42*
Müller, T. *39–42*
Murakami, T. *141*
Muranaka, T. *132*
Murata, E. *137*
Musicki, B. *38*
Musienko, M.M. *139*
Myers, T.G. *135*

Nafady, A.M. *141*
Naito, S. *136*
Nakamura, H. *38*
Nakanishi, K. *131*
Nakano, K. *129*
Nakao, Y. *130*
Nanasi, P. *129*
Nandi, O.I. *41*
Navarro, V. *137*
Navarro, Y.M.C. *140*
Nigro, P. *141*
Nikaido, T. *129, 133, 134*
Nimtz, M. *130*
Nishi, M. *140*
Nishida, M. *128*
Nishikawa, H. *136*
Nishimura, K. *141*
Nishino, A. *129, 131*
Nishino, H. *129, 131*
Nisius, A. *130*
Nohara, T. *128, 129, 135, 136, 139, 141*
Noodén, L.A. *43*
Nord, L.I. *133*
Nugent, S. *127*

Oberhuber, M. *40, 42*
Oboshi, S. *136*
Ocampo, R. *43*
O'Connor, P.M. *135*
Ogihara, Y. *141*
Oh, S.-R. *136*
Ohta, H. *43*
Ohtani, K. *127, 131, 137*
Ohtsuki, T. *128, 133*
Ojika, M. *131*
Oka, K. *136*
Okada, K. *39*
Okawa, K. *43*
Okawa, M. *129, 139, 141*
Okihara, M. *131*
Okubo, K. *135, 137*
Oleszek, W. *139, 141*
Ongania, K.H. *39–41*
Ono, M. *129, 141*
Ori, K. *140*
O'Rourke, K. *136*
Ortiz de Montellano, P.R. *38*
Osbourn, A.E. *138*
Oshimura, Y. *131*
Oshio, Y. *42*
Osorio, J.N. *140*
Ougham, H. *38, 43*
Ozipek, M. *141*

Paczkowski, C. *139*
Padron, J.I. *140*
Pal, B.C. *128*
Palatnik, M. *138*
Parente, J.E. *138*
Parente, J.P. *138, 139*
Park, J.H. *131*
Park, M.H. *131*
Pasehnichenko, V.A. *138*
Pathak, A.K. *128*
Paull, K.D. *135, 136*
Pecanha, L.M.T. *138*
Pedersen, C. *134*
Pedersen, H. *134*
Pegg, D.T. *133*
Peisker, C. *38–40*
Pellatin, L.D. *138*
Peng, J.P. *128*
Perreault, H. *132*
Perrone, A. *141*
Petit, P. *127*

Pettini, J.L. *138*
Pettit, G.R. *128*
Pettit, R.K. *128*
Pezzuto, J.M. *139*
Piacente, S. *139, 141*
Pilipenko, V.V. *132*
Pinilla, V. *128*
Piskarev, V. *132*
Pittayakhachonwut, D. *130*
Pizza, C. *131, 139, 141*
Pkheidze, T.A. *127*
Plaza, A. *141*
Plock, A. *130, 134*
Plowman, J. *135*
Pocsi, I. *129*
Poirier, G. *136*
Porra, R. *38*
Poulton, J.E. *129*
Presber, W. *134*
Prieto, J.M. *140*
Pružinska, A. *39–41*
Putalun, W. *132, 141*

Qiu, F. *140*
Quan, L.T. *136*
Qui, Y.L. *41*
Quintana, J. *136, 140*

Radcliffe, E.B. *138*
Rahalison, L. *137*
Rahman, S.K. *127*
Raman, K.V. *138*
Ramberg, J. *127*
Rance, M. *134*
Rao, G.H.R. *134*
Ravelo, A.G. *141*
Raven, H.P. *43*
Regli, P. *137*
Rentsch, D. *38–40*
Rentsch, G.H. *141*
Reynolds, W.F. *140*
Ribes, G. *127*
Riov, J. *43*
Roberts, L. *38*
Roca, M. *41*
Rodoni, S. *41–43*
Rodriguez, S. *131*
Roepstorff, P. *132*
Rokem, J.S. *138*
Roussakis, C. *134*

Rubinstein, L.   *135*, *136*
Rüdiger, W.   38, 39, *130*
Ryu, M.Y.   *136*

Sahu, N.P.   *127–129*, *131*, *133*, *134*, *138*
Saiki, I.   *136*
Saito, E.   *129*
Saito, T.   *136*
Sajjadi, S.E.   *139*
Sakagami, H.   *135*, *136*, *139*, *140*
Sakai, S.   *128*
Sakuma, C.   *135*
Salehpoour, M.   *132*
Salvesen, G.S.   *136*
Sang, S.   *133*
Santokaran, S.   *131*
Santos, W.R.   *138*
Saracoglu, I.   *141*
Sargent, J.M.   63, *135*
Sashida, Y.   *127–131*, *133*, *135–137*, *139–141*
Sata, N.   *136*
Sato, M.   *133*
Satomi, Y.   *129*, *131*
Satou, T.   *140*
Sausville, E.A.   *135*
Sautour, M.   48, *128*, *137*
Save, G.   *132*
Savoy, Y.E.   *138*
Saxena, V.K.   *141*
Scheer, H.   38–40
Schellenberg, M.   38–43
Schenk, N.   41
Schettino, O.   *131*
Scheumann, V.   39
Schiff, J.A.   39
Schlösser, E.   *138*
Schmidt, J.M.   *128*
Schneider-Poetsch, H.A.   *130*
Schoch, S.   39
Schönbeck, F.   *138*
Schulten, H.R.   *132*
Schweigerer, L.   *132*
Scudiero, D.A.   *135*, *136*
Seike, H.   *129*
Selitrennikoff, M.C.P.   *137*
Sen, S.   *135*
Seo, S.   *131*
Seong, J.-D.   *136*
Seong, N.-S.   *136*
Shalem, A.   *141*

Shen, P.   *141*
Shen, Q.   *139*
Shen, X.   *132*
Shevchenko, A.   *132*
Shi, J.-G.   *133*
Shibuya, N.A.   *130*
Shimada, H.   *43*
Shimokawa, K.   39
Shimomura, O.   38, *43*
Shimoyamada, M.   *137*
Shioi, Y.   30, *39*, *42*
Shirley, N.J.   *130*
Shoemaker, R.H.   *135*, *136*
Shoyama, Y.   *132*, *140*, *141*
Siegelman, H.W.   39
Silva, B.P.D.   *138*
Silva, G.D.M.   *138*, *139*
Simon, R.M.   *136*
Singh, S.B.   *127*
Sitek, M.   *141*
Sjolander, A.   *135*
Skaric, V.   38
Skehan, P.   *135*, *136*
Smart, C.M.   *43*
Smith, F.   *130*
Smith, K.M.   38
Soejarto, D.D.   *139*
Sokolowska, W.   *134*
Solomos, T.   39
Son, K.H.   *130*
Song, F.R.   *132*
Song, K.-S.   *136*
Sørensen, O.W.   *134*
Soulé, S.   *138*
Southavong, B.   *139*
Spremulli, L.   39
Srisomsap, C.   *130*
Stahl, E.   *128*
Stamova, A.I.   *130*
Stead, A.D.   *43*
Stephens, R.L.   *134*
Stochmal, A.   *141*
Stocker, R.   *43*
Stojanovic, M.N.   *43*
Stoll, A.   6, *39*
Stone, B.A.   *130*
Stone, M.J.   *127*
Stothers, J.B.   *131*
Sudo, K.   *137*
Sue, M.   *129*

Sukhodub, L.F. *132*
Summers, M.F. *134*
Sun, Q. *128*
Sun, W.X. *132*
Sundqvist, B. *132*
Surarit, R. *130*
Suter, D. *134*
Suzuki, H. *127*
Suzuki, K. *141*
Suzuki, M. *137*
Suzuki, Y. *42*
Suzuo, M. *131*
Svasti, J. *130*
Syahrani, A. *139*
Sydara, K. *139*

Takaashi, Y. *139, 140*
Takamiya, K. *39*
Takamiya, K.-I. *43*
Takamura, S. *136*
Takashi, T. *137*
Takechi, M. *137*
Takeda, K. 46, *127*
Takeda, R. *131*
Takeya, K. *133, 139*
Tal, B. *138*
Tamir, J. *138*
Tamura, Y. *127, 137*
Tan, C.-H. *141*
Tan, G.T. *139*
Tan, N.-H. *140*
Tanaka, A. *39*
Tanaka, H. *132, 140, 141*
Tanaka, K. *136*
Tanaka, N.K. *39*
Tanaka, O. *127, 131, 137*
Tanaka, R. *39*
Tanaka, Y. *39, 137*
Tanner, G. *40, 41*
Tanticharoen, M. *130*
Tapernoux-Lüthi, E. *41*
Tatsumi, Y. *39*
Taylor, C.G. 63, *135*
Taylor, L.C.E. *131*
Techasakul, S. *130*
Temraz, A. *141*
Tewari, M. *136*
Tezuka, Y. *136*
Thakur, R.S. *127*
Thebtaranonth, Y. *130*

Thieulant, M.-L. *134*
Thomas, A. *38*
Thomas, H. *37–43*
Thompson, J. *136*
Tinto, W.F. *140*
Tomè, F. *140*
Tomimatsu, T. *129*
Tomita, Y. *131*
Topalov, G. *41*
Tori, K. *131*
Tortoriello, J. *137*
Tosini, S. *136*
Tran, Q.K. *136*
Tran, Q.L. *136*
Treibs, A. *43*
Troxler, R.F. *38*
Tschesche, R. 46, 69, *127, 138*
Tsuchiya, T. *43*
Tsuji, H. *39*
Tsukamoto, S. *130*
Tsumagari, H. *136, 139*
Turner, A.B. *130*
Turner, D.L. *128*

Ulrich, M. *40*
Unrau, A.M. *130*

Vaigro-Wolff, A. *135*
Van-Binst, G. *131*
Van DeWerken, G. *132*
Van Osdol, W.W. *135*
Van Setten, D.C. *132*
Vasil'eva, I.S. *138*
Vázquez, A. *138*
Vazquez, J.T. *131*
Vicentini, F. *39, 41*
Vigo, C.B. *137*
Vistica, D. *135*
Viswanadhan, V.N. *135*
Vogt, E. *42*
Voigt, G. *127*
Vollerner, Y.S. *130*
Vuister, G.W. *134*

Wagner, G. *134*
Wagner, H. *138*
Wagstaff, C. *43*
Wall, M.E. *129*
Wang, B.-X. *140*
Wang, G. *136*

Wang, J. *130*
Wang, S.-L. *141*
Warren, C.D. *134*
Watanabe, K. *39*, *135*, *136*, *140*
Weeden, N. *38*
Wegner, C. *139*
Weinstein, J.D. *38*
Weinstein, J.N. *135*
Wen, H. *137*
Wiertz, E.J.H.J. *132*
Wiesler, W.T. *131*
Wilkins, A.L. *139*
Wilkins, R.W. *138*
Williams, D.H. *127*, *131*
Willstätter, R. *39*
Wilm, M. *132*
Wink, M. *138*
Wittes, R.E. *135*
Wojciechowski, Z.A. *138*, *139*
Wolfender, J.-L. *131*
Wolters, B. *137*
Woodward, R.B. *38*
Wray, V. *130*
Wu, L.-J. *140*
Wu, Z. *133*
Wulff, G. *46*, *127*
Wüthrich, K. *134*
Wüthrich, K.L. *41*

Xiao, Z. *127*
Xu, M. *140*
Xu, Y. *137*
Xu, Y.-X. *141*
Xuan, L.-J. *141*

Yamaguchi, N. *128*
Yamamoto, Y. *43*
Yamasaki, K. *127*, *130*, *131*, *137*
Yamauchi, T. *137*
Yan, W. *131*
Yan, X. *128*
Yanai, Y. *129*, *141*
Yang, C.-R. *67*, *128*, *130*, *133*, *137*, *139*–*141*
Yang, D.-J. *140*
Yang, Q.-X. *140*, *141*
Yang, Y. *128*
Yang, Y.-C. *133*

Yang, Z. *127*
Yao, X. *135*, *136*, *140*
Yao, X.S. *128*
Yao, X.-S. *141*
Yasumoto, K. *136*
Ye, W. *136*
Yokosuka, A. *50*, *127*, *129*, *133*, *135*, *136*, *139*–*141*
Yong, J. *128*
Yoon, K.-D. *136*
Yoshida, K. *39*
Yoshihara, O. *141*
Yoshikawa, M. *141*
Yoshiki, Y. *135*
Yoshimitsu, H. *128*, *141*
Yoshimura, Y. *131*
Youn, J.-Y. *40*
Yu, B. *128*, *132*
Yu, D.-Q. *139*
Yu, H. *128*
Yves, S. *127*

Zaharevitz, D.W. *135*
Zamilpa, A. *137*
Zampella, A. *134*
Zerbe, O. *134*
Zhang, H.-J. *139*
Zhang, J. *49*, *128*, *137*
Zhang, J.B. *128*
Zhang, Q. *128*
Zhang, Y. *137*
Zhang, Y.-J. *133*, *137*, *139*–*141*
Zhao, Y. *128*
Zheng, Q.-A. *140*
Zhou, G. *136*
Zhou, J. *140*
Zhou, X. *136*
Zhou, Y. *133*
Zhou, Z.L. *131*
Zhu, D.-Y. *141*
Zhu, X.-H. *141*
Zimowski, J. *139*
Zolfaghari, B. *139*
Zomer, G. *132*
Zou, Z.-M. *139*
Zubarev, R.A. *132*
Zumaroh, S. *139*

Subject Index

Abscissic acid 35, 36
Abutiloside L 49, 108
Abutiloside M 49, 108
Abutiloside N 49, 108
Abutiloside O 49
Accelerated cell death genes 36
Acetone 49
(23*S*, 24*S*)-21-Acetoxy-spirosta-5,25(27)-diene-1β,3β,23,24-tetrol 94, 95
(23*S*, 24*R*, 25*R*)-1β-Acetoxy-spirost-5-ene-3β,23,24-triol 102
Acid hydrolysis 50, 53
Aculeatiside A 49, 55, 120
Aculeatiside B 49
Acyclovir 68
Adriamycin 65
Aflagellated ovoid shape 69
Agamenoside H 71
Agamenoside I 71
Agamenoside J 71
Agavaceae 71, 85, 89, 93, 99, 106, 120
Agave americana 71
Agave attenuata 68, 71
Agave brittoniana 72
Agave decipiens 72
Agave fourcroydes 47, 73
Agave shrevei 73
Agave sp. 68
Agavegenin C 71, 122
Agigenin 73, 74, 122
Alditol acetates 50
Alliaceae 74, 76, 94
Allium ampleoprasum 73
Allium elburzense 74
Allium jesdianum 63, 75
Allium karataviense 64, 75
Allium nutans 76
Allium porrum 76
Allium tuberosum 65, 76, 77
Allium vineale 77
Almond emulsin 52
4-^{14}C-δ-Aminolevulinic acid 5

Ammonia 51
Amphotericin B 67
Anguivioside A 109
Anguivioside B 109
Anguivioside C 109
Anguivioside III 108
Anguivioside XI 109
Anisaldehyde 51
Annexin V 69
Antideteriorating activity 67
Antifungal activity 48, 66–68, 126
Antileishmanial activity 68, 69
Antineoplastic activity 48
Antioxidant activity 29, 36
Antiproliferative activity 66
Antiviral activity 68
Antiyeast activity 67
Aphids 68
Apiose 51, 125
D-Apiose 50
Apoptosis 66
Apples 29
Aqueous acid 14, 15
Arabidopsis thaliana 18, 24, 27, 28, 36
Arabinose 125
L-Arabinose 50
Aspafilioside D 78
Aspaoligonin A 65, 79
Aspaoligonin B 65, 79
Aspaoligonin C 65
Asparacoside 1 78
Asparagaceae 78
Asparagus africanus 78
Asparagus cochinchinensis 56, 78
Asparagus filicinus 78
Asparagus officinalis 79
Asparagus oligoclonos 65, 79
Asparagus racemosus 47, 61, 68, 79
Aspergillus fumigatus 67
Aspergillus niger 67
Atmospheric pressure chemical ionization mass spectrometry 55

Atropuroside A   107
Atropuroside B   67, 107
Atropuroside C   107
Atropuroside D   107
Atropuroside E   107
Atropuroside F   67, 107
Atropuroside G   107

Bacteriochlorophyll c   15
Bacteriochlorophylls   32
Baeyer-Villiger oxidation   50
Balanites aegyptica   80
Barium chloride   51
Barley   5, 7, 8, 11, 26–28, 30
(24S, 25S)-3-O-Benzoyl-5α-spirostane-2α,3β,5,6β,24-pentol   75
Bilirubin   21, 36
Biliverdin   20, 21, 36
Biological activity   46, 62
Bisdesmosidic cholestane glycoside   64
Brassica napus   11, 17, 26
Breast cancer   64
Breast cancer MCF-7   63, 65
Breast carcinoma   65
Bufadienolides   69
Bulbus Allii Macrostemi   94
n-Butanol   47–49

Calamus insignis   80
Camassia leichtlinii   81
Cancer cells   126
Candida albicans   67
Candida glabrata   67
Candida kefyr   67
Candida krusei   67
Candida tropicalis   67
Canola sp.   16, 26
Capsicum annuum   12, 24
Carboplatin   65
Carboxylic acid   10, 27
13²-Carboxy-pyropheophorbide a   9, 10, 28
Cardenolides   69
Cardiac glycosides   45, 46
Caspase   66
CDDP   66
Cell growth inhibitory activity   65
Cell shrinkage   69
Central nervous system carcinoma   65
Cercidiphyllum japonicum   4, 24, 27
Cestrum nocturnum   65, 83

Cestrum sendtenerianum   84
Chacotriose   125
Chacotriose derivatives   65
Chenopodium album   8–10, 28, 31
Chlamydomonas reinhardtii   9
Chl a   2, 3, 5–8, 27, 28, 30, 31
Chl b   2, 3, 7, 8, 27, 28, 30, 31
Chlorella protothecoides   9, 10, 15, 18, 30–32
Chlorella sp.   32
Chlorins   5, 6, 8, 10, 13, 15, 35, 36
Chloroform   47, 49
Chlorogenin   67, 83, 101, 109, 122
β-Chlorogenin   73, 76, 78, 122
Chlorophyll   2, 8, 28, 32, 35, 36
Chlorophyll a   2, 3, 7, 11
Chlorophyll b   2, 3, 7
Chlorophyllase   6, 7, 36
Chlorophyll catabolites   2, 33
Chlorophyllide a   6–8, 11
Chlorophyllide b   6, 8
Chlorophylls   35
Chlorophytum comosum   51
Cholesterol   69
Chromatin condensation   69
Chromatography   47, 48, 51
Chrysogenin   109, 122
CNS cancer   64
Collision-induced decomposition   55, 56
Colon cancer   64, 65
Colon 26-L5 carcinoma   66
Column chromatography   49
Convallariaceae   101, 107, 118
Convallogenin B   119, 122
Cordyline stricta   85
Corn   27
Costaceae   86
Costus speciosus   52
Costus spicatus   86
Crape ginger   69
Crestagenin   77, 122
Cryptococcus neoformans   67
Cucurbitaceae   94
Cyanogenic glucosides   51
Cycloartenol   69
(24S, 25R)-3α,5α-Cyclospirostane-1β,6β,24-triol   93
3,5-Cyclospirostanol   50
Cytokinin   36
Cytostatic activity   63, 64
Cytotoxic activity   63–66, 126

Subject Index

Death genes 18
17^2-Decarboxy-13^1-deoxophyto-porphyrinate 35
Degalactotigonin 65
9,11-Dehydromanogenin 63, 96, 122
13^4-Demethyl-pFCC 29
2-Deoxyribose 50
21-Deoxytrillenogenin 118, 125
Deoxytrillenoside B 118
Detoxification 36
Deuterium 32
Develosil ODS HG-5 48
Diaion HP-20 47
$3^1,3^2$-Didehydro-4,5,10,17,18-(22H)-hexahydro-13^2-(methoxycarbonyl)-4,5-dioxo-4,5-seco-phytoporphyrin 16
$3^1,3^2$-Didehydro-1,4,5,10,17,18,20-(22H)-octahydro-13^2-(carboxy)-4,5-dioxo-4,5-seco-phytoporphyrin 27, 28
$3^1,3^2$-Didehydro-1,4,5,10,17,18,20-(22H)-octahydro-13^2-(methoxycarbonyl)-4,5-dioxo-4,5-seco-phytoporphyrin 11, 13, 17, 18
$3^1,3^2$-Didehydro-1,4,5,10,15,20-(22H,24H)-octahydro-13^2-(methoxycarbonyl)-4,5-dioxo-4,5-seco-phytoporphyrinate 24
Diethyl ether 49
15,16-Dihydrobiliverdin 20, 21
Dihydrogen 18
(3E)-2,3^2-Dihydro-RCC methyl ester 19
Dilute acid 8
p-Dimethylaminobenzaldehyde 50
3-(4,5-Dimethylthiazol-2-yl)-2,5-diphenyl tetrazolium bromide 65
Dioscin 65, 67
Dioscorea cayenensis 86
Dioscorea collettii var. *hypoglauca* 64
Dioscorea floribunda 47
Dioscorea panthaica 55, 86, 87
Dioscorea polygonoides 87
Dioscorea pseudojaponica 87
Dioscoreaceae 86
Dioscoreside A 86
Dioscoreside B 87
Dioscoreside C 87
Dioscoreside D 87
Diosgenin 66, 67, 76, 80, 88, 97, 107, 122
Diosgenin-rhamno-glucoside 64
Diosgenone 66
Dioxane 51

1,20-Dioxo-1,20-secopheophorbidates 15
1,20-Dioxo-1,20-secophytoporphyrinate 34
1,20-Dioxo-1,20-secopyropheophorbides 32
Dioxygen 18
Disporopsis pernyi 88
Disporoside A 88
Disporoside B 88
Disporoside C 88
Disporoside D 88
Doxorubicin 65, 66
Dracaena angustifolia 66, 88
Dracaena cochinchinensis 89
Dracaena concinna 90
Dracaena draco 64, 91
Dracaena surculosa 92
Dracaenaceae 88
Dracaenoside I 89
Dracaenoside J 90
Dracaenoside K 90
Dracaenoside L 90
Dracaenoside R 90
Draconin A 91
Draconin B 91
Draconin C 91

Ehrlich reagent 50
Elburzenoside A1 74
Elburzenoside A2 74
Elburzenoside B1 74
Elburzenoside B2 74
Elburzenoside C1 74
Elburzenoside C2 74
Elburzenoside D1 75
Elburzenoside D2 75
Electrophilic agents 5
Electrospray ionization 55, 126
Electrospray ionization mass spectrometry 56
Enzymatic activity 12
Enzymatic oxygenating activity 11
Epiyamogenin 122
22-Epiyamogenin 81
(22S, 23S, 25R, 26S)-23,26-Epoxy-5α-furostane-3β,22,26-triol 71
22,25-Epoxy-furost-5-ene 49
(22S, 25S)-22,25-Epoxy-furost-5-ene-3β,14α,26,27-tetrol 90
(22S, 25S)-22,25-Epoxy-furost-5-ene-3β,7β,26-triol 108
(22S, 25S)-22,25-Epoxy-7β-methoxy-furost-5-ene-3β,26-diol 108

Ethanol 46, 49
Ethyl acetate 46–48
Ethyl-methyl-maleimide 30
Etoposide 63, 64
Euphasia pacifica 32

Fast atom bombardment 56
Bn-FCC-2 11, 12, 17
Ca-FCC-2 12, 23
FCCs 10, 11, 17, 21, 22, 24, 27, 28
Ferredoxin 17, 19, 20
Festuca pratensis 5, 13
Fibrosarcoma HT-1080 66
Floribundasaponin A 47
Floribundasaponin B 47
Floribundasaponin C 47
Floribundasaponin D 47
Floribundasaponin E 47
Floribundasaponin F 47
Fluorescent chlorophyll catabolites 22
Fluorescing Chl-catabolites 11
French beans 28
Fructus Trichosanthis 94
FT-NMR 52
Fucose 125
L-Fucose 50
Furcraea selloa var. *marginata* 93
Furcrea furostatin 93
Furcrea furostatin methyl ether 93
(22S, 25S)-Furospirostane-1α,2β,3α,5α,26-pentol 119
(25S)-Furosta-5,20(22)-diene-3β,26-diol 113
Furosta-5,25(27)-diene-22ξ-methoxy-1β,2α,3β,26-tetrol 107
Furosta-5,25(27)-diene-1β,2α,3β,22ξ,26-pentol 107
(25R)-Furosta-5,20(22)-diene-1β,3β,26-triol 89
(25R)-Furosta-5,20(22)-diene-2α,3β,26-triol 83
20,22-*seco*-Furosta-5,25(27)-diene-1β,3β,26-triol-20,22-dione 89
Furostane analogues 58
Furostane glycosides 65
Furostane-2α,3β,5α,6β,22α,26-hexol 74
Furostane-2α,3β,5α,6β,22β,26-hexol 74
Furostane-2α,3β,5α,22α,26-pentol 74, 75
Furostane-2α,3β,5α,22β,26-pentol 74, 75
(25S)-Furostane-3β,5β,6α,22ξ,26-pentol 77

Furostanes 46
Furostane steroids 57
(25R)-5α-Furostane-2α,3β,22α,26-tetrol 115
(25R)-5α-Furostane-2α,3β,22ξ,26-tetrol 117, 118
(25S)-5α-Furostane-2α,3β,22ξ,26-tetrol 117
(25R)-5α-Furostane-3β,6β,22ξ,26-tetrol 77
(25R)-5α-Furostane-3β,22α,26-triol 115
(25S)-5α-Furostane-3β,22α,26-triol 114, 117
(25R)-5α-Furostane-3β,22ξ,26-triol 77, 93
(25R)-5β-Furostane-3β,22α,26-triol 120
(25S)-5β-Furostane-3β,22,26-triol 78
(22R, 25R)-5β-Furostane-3β,22,26-triol 88
(25R)-5α-Furostane-3β,22α,26-triol-12-one 100, 101
Furostanol 50
Furostanol glucosides 51
Furostanol glycosides 49, 50–52, 65, 66
Furostanol saponins 49, 56, 63
(25S)-Furost-4,20(22)-diene-26-ol-3,12-dione 117
Furost-5,25(27)-diene-1β,3α,22ξ,26-tetrol 119
(25S)-5β-Furost-20(22)-ene-3β,26-diol 116
(25R)-20,22-*seco*-Furost-5-ene-3β,26-diol 20,22-dione 86, 87
(25R)-5β-Furost-20(22)-ene-3β,26-diol-12-one 121
5β-Furost-25(27)-ene-1β,3β,6β,22α,26-pentol 94
(22R, 23S, 25R, 26S)-Furost-5-en-23,26-epoxide-3β,22α,26-triol 109
Furost-5-ene-1β,3α,22,26-tetrol 119
(25R)-Furost-5-ene-1β,3β,22ξ,26-tetrol 101, 102
(25S)-Furost-5-ene-1β,3β,22ξ,26-tetrol 102
(20R, 25S)-5α-Furost-22-ene-2α,3β,20,26-tetrol 77
(20S, 25S)-5α-Furost-22-ene-2α,3β,20,26-tetrol 77
(25R)-Furost-4-ene-3β,22ξ,26-triol 118
(25S)-Furost-4-ene-3β,22ξ,26-triol 118
(25R)-Furost-5-ene-3β,22α,26-triol 81
(25S)-Furost-5-ene-3β,22ξ,26-triol 118
(22α, 25R)-Furost-5-ene-3β,22,26-triol 117
(20S, 25S)-5α-Furost-22-ene-3β,20,26-triol 77
(23S, 25R)-20,22-*seco*-Furost-5-ene-3β,23,26-triol-20,22-dione 87

Subject Index

(25R)-Furost-5-ene-3β,22ξ,26-triol-12-one 102
(25S)-Furost-5-ene-3β,22ξ,26-triol-12-one 101, 102

Galactose 125
D-Galactose 50, 53
Garlic 68
Geo-porphinoids 35
Gitogenin 63, 120, 122
Gitogenin diglycoside 63
Glioma 65
Gloriogenin 78, 122
β-Glucopyranosyl 26
Glucose 125
D-Glucose 50
26-O-β-Glucosidase 52
β-Glucosidases 51, 52
Glucosides 62
26-Glucosyloxyfurostanol saponin 63
Glycoalkaloids 70
Glycosides 45
Gracillin 64
Green algae 9, 10, 15, 18, 30, 31, 32, 34
Green gene 36
Gulose 125

Hapten 55
Hecogenin 67, 101, 116, 122
Helleborus orientalis 94
Helleborus viridis L. 95
Hematinic acid 30
Hemerocallis furva var. *kwanso* 95
Hemeroside A 95
Hemeroside B 95
Herpes simplex virus type 1 68
Hesperidinase 52
Heteronuclear multibond correlation 62
n-Hexane 48, 49
Hexasaccharides 47, 48
High-resolution mass spectrometry 11
Hordeum vulgare 26
Hosta sieboldii 63, 96
HPL-chromatography 11, 47–50, 52, 55, 126
Hydrochloric acid 50
Hydrolysis 52, 54
Hydro-porphinoids 22
Hydroxamic acid glucosides 51
(25R)-3-O-(2-Hydroxybutyryl)-5α-spirostane-2α,3β,5,6β-tetrol 75

10-Hydroxycamptothecin 65
1β-Hydroxy-crabbogenin 85, 122
(24S)-Hydroxyneotokorogenin 95, 122
(25R,22ξ)-Hydroxywattinoside C 101
Hypocholesterolaemic effects 68

Icogenin 66, 92
Ion trap tandem mass spectroscopy 55, 56
Isoflavonoid glucosides 51
Isonarthogenin 97, 98, 103, 122
Isonuatigenin 111, 122
Isopentenyl pyrophosphate 69
Isorhodeasapogenin 95, 122
Isoterrestrosin B 64, 116

Kallstroemia pubescens 47
Kallstroemin A 47
Kallstroemin B 47
Kallstroemin C 47
Kallstroemin D 47
Kallstroemin E 47
Karplus relationship 62
Katsura tree 4
β-Ketocarboxylic acid 10, 27, 28
Ketoconazole 67
(25S)-Kingianoside C 101
(25S)-Kingianoside D 101
Kingianoside E 102
(25S)-Kingianoside E 102
Kingianoside F 102
(25S)-Kingianoside F 102
Klyne's rule 52
Krill 5, 32

Leguminosae 117
Leishmania donovani 68, 69
Leishmaniasis 68
Leukemia 64
Leukemia HL-60 64–66
Liliaceae 50, 73, 75, 76, 78, 81, 88, 96–98, 103
Lilium candidum 97
β-Linked oligoglucosides 51
Liquidambar styraciflua 27
Liver carcinoma 65
LOGIT method 65
Luciamin 68, 110
Luciferase 32
Luciferin 5, 32–34
Lung cancer 66
Lung cancer HOP-62 63

Lung carcinoma 65
β-Lycotetraosyl spirostanol 65

Macranthogenin 121, 122
Magnesium 8, 36
Malignant melanoma 65
Malonic acid 26
Malonyl 26
Manogenin 63, 96, 122
Marine organisms 32–34
Marker's degradation 50
Mass spectrometry 12, 17, 20, 26, 28, 31, 46, 55, 56, 126
Melanoma 64
Melanoma B-16 BL6 66
Metabolic channeling 18
Methanol 20, 46–51
Methanolysis 52
Methotrexate 63
(25S)-22α-Methoxy-3α,5α-cyclofurostane-1β,6β,26-triol 93
(23S, 25R)-23-Methoxy-furosta-5,20(22)-diene-3β,26-diol 87
22ξ-Methoxy-furosta-5,25(27)-diene-1β,3β,26-triol 72, 86, 103, 104
(25R)-22α-Methoxy-5α-furostane-3β,26-diol 115, 116
(25R)-22ξ-Methoxy-5α-furostane-3β,26-diol 73, 83
(25R)-22α-Methoxy-5β-furostane-3β,26-diol 78
(25S)-22α-Methoxy-5β-furostane-3β,26-diol 71, 108
(25R)-22ξ-Methoxy-5α-furostane-2α,3β,5,6β,26-pentol 76
(25R)-22α-Methoxy-5α-furostane-2α,3β,26-triol 115
(25R)-22-Methoxy-5α-furostane-3β,22ξ,26-triol 93
(25R)-22α-Methoxy-5α-furostane-2α,3β,26-triol-12-one 96
(25R)-22α-Methoxy-furost-5-ene-3β,26-diol 72, 86, 87, 95, 117
(25S)-22α-Methoxy-furost-5-ene-3β,26-diol 112
(25R)-22ξ-Methoxy-furost-5-ene-3β,26-diol 86, 97
(25S)-22ξ-Methoxy-furost-5-ene-3β,26-diol 92
(20S, 22R, 25R)-22-Methoxy-furost-5-ene-3β,26-diol 80

(20S, 22R, 25S)-22-Methoxy-furost-5-ene-3,26-diol 80
22ξ-Methoxy-furost-25(27)-ene-1β,2β,3β,4β,5β,7α,26-heptol-6-one 119
22ξ-Methoxy-5α-furost-25(27)-ene-1β,3α,4α,26-tetrol 90
22ξ-Methoxy-5α-furost-25(27)-ene-1β,3β,4α,26-tetrol 91
(20R, 25S)-20-Methoxy-5α-furost-22-ene-2α,3β,26-triol 76
22ξ-Methoxy-5α-furost-25(27)-ene-1β,3α,26-triol 90
22ξ-Methoxy-5α-furost-25(27)-ene-1β,3β,26-triol 85
(25R)-22ξ-Methoxy-furost-5-ene-1β,3β,26-triol 105, 106
(25S)-22α-Methoxy-furost-5-ene-1β,3β,26-triol 92
(25R)-22α-Methoxy-furost-5-ene-2α,3β,26-triol 83
(25R)-22α-Methoxy-5α-furost-9-ene-2α,3β,26-triol-12-one 96
20-Methoxy-pyropheophorbidate 34
(25R, 26R)-26-Methoxy-spirost-5-ene-3β-diol 97
(25R, 26R)-26-Methoxy-spirost-5-ene-17α,3β-diol 97, 98
(25R, 26R)-26-Methoxy-spirost-5-en-3β-ol 111
Methyl-3^1-dehydro-2,4,5,10,17,18,22-heptahydro-13^2-(methoxycarbonyl)-4,5-dioxo-4,5-seco-(22H)-phytoporphyrin 20
Methyl-3^1,3^2-didehydro-1,4,5,10,17,18,20,22-octahydro-13^2-(methoxycarbonyl)-4,5-dioxo-4,5-seco-(22H)-phytoporphyrin 20
Methyl 3,6-di-O-methyl-D-galactopyranoside 53
Methyl 4,6-di-O-methyl-D-glucopyranoside 53
4-Methyl-2,5-dioxo-2,5-dihydropyrrole-3-propionic acid 30
Methyl-4,5-dioxo-4,5-secopheophorbidate a 14, 15
Cd-Methyl-4,5-dioxo-4,5-secopheophorbidate a 14, 15
Cd-Methyl-19,20-dioxo-19,20-secopheophorbidate a 14, 15

Subject Index

Zn-Methyl-19,20-dioxo-19,20-
  secopheophorbidate *a*   14, 15
Methylene chloride   48
Methyl ester   28
Methyl glycosides   60
22-*O*-Methyl-parvispinoside A   115
22-*O*-Methyl-parvispinoside B   115
Methyl-pheophorbidate *a*   14, 15
Cd-Methyl-pheophorbidate *a*   14, 15
Zn-Methyl-pheophorbidate *a*   14, 15
Methyl protogracillin   64
Methyl protoneogracillin   64
Methyl prototribestin   117
Methyl 2,3,4-tri-*O*-methyl-
  L-rhamnopyranoside   53
Methyl 2,3,4-tri-*O*-methyl-D-xylopyranoside
  53
Microhydrolysis   50
Mimusopin   62
Mimusops elengi   62
Mono-oxygenase   31
Monosaccharides   51, 52
MTT dye-reduction assay method   66
Mulifidoside   120
Multi-stage tandem mass spectrometry   55
Murine leukemia P388   65

Namonin A   88
Namonin B   89
Namonin C   89
Namonin D   89
Namonin E   89
Namonin F   89
At-NCC-1   25
At-NCC-2   25
At-NCC-3   25, 27
At-NCC-4   25
At-NCC-5   25
Bn-NCC-1   25, 26
Bn-NCC-2   25, 26
Bn-NCC-3   25, 26
Bn-NCC-4   25, 26
Bn-NCCs   26
Cj-NCC-1   24–27
Cj-NCC-2   23–25
Hv-NCC-1   3, 5, 6, 8, 21, 22, 24–27, 29,
  30
Nr-NCC-1   25
Nr-NCC-2   25
So-NCC-1   25

So-NCC-2   25–27
So-NCC-3   25, 27
So-NCC-4   25, 27
So-NCC-5   25
Zm-NCC-1   25
Zm-NCC-2   25
NCCs   3, 5, 8, 10–12, 22–30, 36
Neochlorogenin   109, 122
Neogitogenin   114, 122
Neohecogenin   67, 73, 81, 122
Neoprotodioscin   115
Neoruscogenin   64, 72, 103, 104, 123
Neosibiricoside A   102
Neosibiricoside B   102
Neosibiricoside C   65, 103
Neosibiricoside D   65
Neotigogenin   53, 54, 67, 115, 122
Nickel-porphyrinate   35
Nitrogen   36, 46
NMR spectroscopy   20, 46, 52, 55, 57, 126
^{13}C NMR spectroscopy   58
2D NMR spectroscopy   55, 59, 126
^{1}H NMR spectroscopy   57
NOE measurements   60
Non-fluorescent chlorophyll catabolites   3,
  5, 22, 25, 27
18-Norspirostanol derivatives   50
Nuatigenin   120, 124

ODS column chromatography   47, 48
Oilseed rape   8, 11, 12, 17, 18, 26, 27, 28
Oligosaccharides   46, 51
Ophiojaponin C   98, 122
Ophiopogenin   98
Ophiopogon japonicus   98
Oral squamous cell carcinoma   65
Ornithogalum thyrsoides   98
Ornithosaponin A   99
Ornithosaponin B   99
Ornithosaponin C   99
Ornithosaponin D   99
Osmium tetroxide   27
Ovarian cancer   64
Ovary malignant ascites   65
2,3-Oxidosqualene   69

Palmae   80
PaO   13
Paper chromatography   51
Paris quadrifolia   50

Parvispinoside A 115
Parvispinoside B 115
Pear tree 28
Pears 29
Pennogenin 111, 122
Pentologenin 118, 122
Periodate oxidation 52
Permethylation 52
Petroleum 35
Petroleum ether 47
Petro-porphyrins 35
pFCC 12, 13, 16–18, 21, 27, 28
pFCCs 3, 20, 22, 24
epi-pFCC 12
1-*epi*-pFCC 12, 13, 23
pFCC methyl ester 19
1-*epi*-pFCC-methyl ester 19
Phaseolus vulgaris 28
Pheo *a* 3, 8, 10–13, 15–18, 28, 30
Pheo *b* 8, 13, 30
Pheophorbide *a* 3, 7–9, 12, 14, 16
Pheophorbide *a* oxygenase 13, 16, 17, 20
Pheophorbides 33
Pheophytin *a* 7
Photooxygenolysis 15
Phycobilins 20, 21
Phytochromobilin 20, 21
(3Z)-Phytochromobilin 21
Phytohormones 35, 36
Phytol 6, 7
Phytyl acetate 7
Pig liver esterase 14, 15
Pink pigments 5
pNCC 23
epi-pNCC 23
Polianthes tuberosa 64, 65, 99
Polianthoside B 99
Polianthoside C 100
Polianthoside D 100
Polianthoside E 100
Polianthoside F 100
Polianthoside G 101
Polycarpon succulentum 51
Polygonatoside A 103
Polygonatoside B 103
Polygonatoside C 103
Polygonatoside D 103
Polygonatum kingianum 49, 101
Polygonatum sibiricum 65, 102
Polygonatum zanlanscianense 103

Porphinoids 2
Porphyrins 35
Porrigenin B 73, 122
Potassium rhodizonate 51
Pregnane glycosides 65
Primary fluorescent chlorophyll catabolites 3, 12, 13, 17
Promyelocytic leukemia HL-60 cells 63, 64
Propionic acid 15, 23, 24
Prosapogenins 51, 53, 54
Prostate cancer 64
Protodioscin 52, 65
Protogracillin 52
Prototribestin 117
Pyrocystis lunula 5, 32
Pyropheo *a* 9, 10, 28
Pyropheophorbide *a* 9, 28

Quantitative fluorescent microscopy 66
Quinovose 67, 125
D-Quinovose 50

Racemoside A 47, 61, 68, 69, 79
Racemoside B 47, 79
Racemoside C 47, 79
Ranunculaceae 94, 95
RCC 3, 13–21
RCC methyl ester 14, 19
RCC-reductase 13, 15, 16, 18–20, 32, 36
Red chlorophyll catabolite 3, 13, 14, 16, 18
Renal cancer 64
Rhamnose 125
Rockogenin 82, 122
ROESY spectrum 61
RP-14 3, 5, 6
Ruscogenin 76, 98, 105, 122
Ruscogenin 1-acetate 102, 122
Ruscogenin diglycoside 63
Ruscus aculeatus 51, 63, 103
Rusty pigment 14 3, 5
Rusty pigments 5

SA III 117
Sansevieria ehrenbergii 48, 106
Sansevierin A 48, 106
Sansevistatin 1 48, 106
Sansevistatin 2 48, 107
Sapogenins 58, 60, 70
Saponin 1 109

Subject Index

Saponin-I 72
Saponin-II 72
Saponin-III 72
Saponin-IV 73
Saponin SC-2 109
Saponin SC-3 109
Saponin SC-4 109
Saponin SC-5 109
Saponin SC-6 109
Saponins 45, 46, 51, 52, 55, 57, 62, 65, 70, 126
Sarsasapogenin 58, 59, 71, 78, 79, 116, 122
Sarsasapogenin M 79
Sarsasapogenin N 79
Sceptrumgenin 89, 106, 123
Schidegeragenin C 94, 121, 122
Schidegera saponin A1 121
Schidegera saponin A2 121
Schidegera saponin A3 121
Schidegera saponin B1 121
Schidegera saponin C1 121
Schidegera saponin C2 121
Scrophulariaceae 120
1,20-Seco-pyropheophorbidate 33
Sephadex LH-20 48
Serum cholesterol 126
Silica gel 5, 47–49, 51
Smilacaceae 108
Smilacina atropurpurea 67, 107
Smilagenin 58, 59, 78, 88, 108, 122
Smilax medica 48, 67, 108
Smith degradation 52
Sodium borohydride 14, 15
Sodium metaperiodate 53
Sodium methoxide 51
Soft ionization mass spectrometry 52, 55
Solakhasoside 1 110
Solanaceae 83, 108, 109
Solanigroside C 110
Solanigroside D 110
Solanigroside E 110
Solanigroside F 111
Solanigroside G 111
Solanigroside H 111
Solanum abutiloides 49, 108
Solanum anguivi 108
Solanum chrysotrichum 67, 109
Solanum hispidum 67, 109
Solanum khasianum 110
Solanum laxum 68, 110

Solanum lyratum 65
Solanum nigrum 65, 110
Solanum sisymbriifolium 111
Solanum sodomaeum 111
Solanum torvum 68, 111
Solatriose 125
Solvolysis 51
Spinach 27, 28
Spirosolane 65
(23S, 24S)-Spirosta-5,25(27)-diene glycoside 64
Spirosta-5,25(27)-diene-1β,2α,3β,12β-tetrol 85
Spirosta-5,25(27)-diene-1β,2α,3β,23α-tetrol 107
(23S, 24R)-Spirosta-5,25(27)-diene-1β,3β,23,24-tetrol 88, 89
(23S, 24S)-Spirosta-5,25(27)-diene-1β,3β,23,24-tetrol 91, 94
Spirosta-5,25(27)-diene-1β,2α,3β-triol 84, 85, 107
(23S)-Spirosta-5,25(27)-diene-1β,3β,23-triol 91, 92, 94, 104, 105
Spirosta-5-ene-3β,14,24-triol 90
Spirostane analogues 58
(25S)-5α-Spirostane-1β,3α-diol 85
(25R)-5α-Spirostane-1β,3β-diol 98
(25R)-5α-Spirostane-3β,15α-diol 81, 82, 111
(25S)-Spirostane-3β,17α-diol 79
(25R)-5α-Spirostane-3β,23α-diol 111
(25R)-5α-Spirostane-3β,6α-diol-12-one 72
(25R)-5α-Spirostane-3β,15α-diol-12-one 82
(25S)-5α-Spirostane-6α,26-diol-3-one 112
(22R, 25R)-5α-Spirostane-3β,23α-diol-26-one 110
Spirostane glucosides 51
Spirostane glycosides 69
(24S, 25S)-Spirostane-1β,2β,3β,4β,5β,7β,24-heptol-6-one 119
(24S, 25S)-5α-Spirostane-2α,3β,5,6β,24-pentol 75
Spirostanes 46, 49, 65
(25R)-5α-Spirostane-1β,2α,3β-triol 85
(25R)-5α-Spirostane-2α,3β,6α-triol 75
(25R)-5α-Spirostane-2α,3β,12β-triol 97
(24S, 25S)-5β-Spirostane-1β,3β,24-triol 119
(22R, 23R, 25S)-5α-Spirostane-3β,6α,23-triol 112

(22*R*, 23*S*, 25*R*)-5α-Spirostane-3β,6α,23-triol 111
(22*R*, 23*S*, 25*S*)-5α-Spirostane-3β,6α,23-triol 111
(22*S*, 23*S*, 24*R*, 25*S*)-5α-Spirostane-3β,23,24-triol 71
(22*R*, 25*R*)-5α-Spirostane-3β,15α,23α-triol-26-one 110
Spirostanol glycosides 64–67
Spirostanol saponins 56, 65, 66
Spirost-5,25(27)-diene-1β,3α,24β-triol 119
(25*R*)-Spirost-5-ene-3β,7α-diol 106
(22*R*, 25*S*)-Spirost-5-ene-3β,15α-diol 110
(23*S*, 25*R*)-Spirost-5-ene-3β,23-diol 97, 106
(24*S*, 25*R*)-Spirost-5-ene-3β,24-diol 113, 114
(25*R*, 26*R*)-Spirost-5-ene-3β,26-diol 109
(25*S*)-Spirost-5-ene-3β,27-diol-12-one 103
(25*R*)-5α-Spirost-9-ene-3β-ol-12-one 101
(25*R*)-Spirost-5-ene-1β,2α,3β,17α-tetrol 107
(23*S*, 25*R*)-Spirost-5-ene-3β,12α,17α,23-tetrol 87
(23*S*, 25*R*)-Spirost-5-ene-3β,14α,17α,23-tetrol 87
Spirost-5-ene-3β,14,27-triol 90
5α-Spirost-25(27)-ene-1β,2α,3β-triol 85
(25*R*)-Spirost-5-ene-1β,2α,3β-triol 84, 107
(25*R*)-Spirost-5-ene-2α,3β,15β-triol 84
(25*R*)-Spirost-5-ene-2α,3β,17α-triol 84
(24*S*, 25*R*)-Spirost-5-ene-1β,3β,24-triol 92
(24*S*, 25*S*)-Spirost-5-ene-1β,3β,24-triol 98, 99
(24*S*, 25*S*)-Spirost-5-ene-2α,3β,24-triol 83
(23*S*, 25*S*)-Spirost-5-ene-3β,17α,23-triol 110
(23*S*, 24*R*, 25*S*)-Spirost-5-ene-3β,23,24-triol 87
(22*R*, 23*S*, 25*R*, 26*R*)-Spirost-5-ene-3β,23,26-triol 108
(23*S*, 25*S*)-Spirost-5-ene-3β,23,27-triol-12-one 103
5β-Spirost-25(27)-en-3β-ol-12-one 121
Squalene 69
Squalene monooxygenase 69
Steroidal glycosides 46, 48, 50, 51, 55, 58, 64, 65, 126
Steroidal hormones 46
Steroidal sapogenins 67, 69
Steroidal saponins 46–50, 56, 58, 60, 62–64, 66–71, 126

Sulfuric acid 51
Surculoside A 92
Surculoside B 92
Surculoside C 92
Surculoside D 92
Sweet gum 27
Sweet pepper 12, 18

T cell lymphoblast-like cell line 63
Tacca chantrieri 65, 112
Taccaceae 112
Tandem mass spectrometry 56
Terrestrinin A 117
Terrestrinin B 117
(23*S*, 24*S*, 25*S*)-1β,3β,23,24-Tetrahydroxyspirost-5-en-15-one 99
Tetrapyrroles 2, 5
Tetrapyrrolic compounds 2
Thin layer chromatography 5, 50
Tigogenin 58, 63, 67, 99, 100, 122
Tigogenin triglycoside 63
Tobacco 27
Torvanol A 68
Torvoside H 68, 112
Torvoside J 111
Torvoside K 111
Torvoside L 112
Toxicity 62
Tribulosaponin A 116
Tribulosaponin B 116
Tribulosin 53, 54
Tribulus alatus 114
Tribulus parvispinus 115
Tribulus terrestris 53, 64, 115
Trigonella foenum-graecum 117
Trigoneoside Xa 117
Trigoneoside Xb 117
Trigoneoside XIb 118
Trigoneoside XIIa 118
Trigoneoside XIIb 118
Trigoneoside XIIIa 118
$3^1,3^2,8^2$-Trihydroxy-1,4,5,10,15,20-(22*H*,24*H*)-octahydro-13^2-(methoxycarbonyl)-4,5-dioxo-4,5-seco-phytoporphyrinate 5, 21
Trillenogenin 118, 125
Trillenoside C 118
Trilliaceae 118
Trillium kamtschaticum 50, 64, 118
Trillium tschonoskii 50

Subject Index

Trisaccharides   47, 48
Triterpenoid saponins   46
Tuberoside   77
Tuberoside F   76
Tuberoside G   77
Tuberoside H   77
Tuberoside I   77
Tumor diseases   64
Tupistra wattii   118
Tupistra yunnanensis   119
Tupistroside A   119
Tupistroside B   119
Tupistroside C   119
Tupistroside D   119
Tupistroside E   119
Tupistroside F   119

Vacuum-liquid chromatography   49
Vanadyl-deoxo-phylloerythroetioporphyrin   35
Vanadyl-porphyrinate   35
Vegetation index   4
Veronica fushii   120
Veronica multifida   120

Water   17, 46–49, 51
Wattoside G   118
Wattoside H   119
Wattoside I   119
Western blot analysis   66

Xylose   53, 67, 125
D-Xylose   50

Yamogenin   69, 80, 81, 103, 113, 114, 117, 122
Yayoisaponin A   65, 73
Yayoisaponin B   65, 73
Yayoisaponin C   65, 74
Yucca filamentosa   120
Yucca gloriosa   67
Yucca schidigera   67, 120
Yuccagenin   84, 122
Yuccaloeside B   67
Yuccaloeside C   67

Zorbax SB C_{18} 48
Zygophyllaceae   80, 114